Pollution in the Electric Power Industry

Pollution in the Electric Power Industry

Its Control and Costs

David L. Scott
Florida Southern College
Lakeland, Florida

Lexington Books
D.C. Heath and Company
Lexington, Massachusetts
Toronto London

Library of Congress Cataloging in Publication Data

Scott, David Logan, 1942-
Pollution in the electric power industry.

Bibliography: p.
1. Electric power-plants–Environmental aspects.
I. Title.
TD195.E4S3 621.312'1 73-6559
ISBN 0-669-89219-X

Published simultaneously in Canada.

Printed in the United States of America.

International Standard Book Number: 0-669-89219-X

Library of Congress Catalog Card Number: 73-6559

To my wife, Kay,
for her patience and help

Contents

List of Figures

List of Tables

Preface

The United States is caught in an energy crisis of which the electric power industry plays a large part. Many areas of the country are experiencing shortages of electricity, and some power company officials contribute a large part of the blame directly to environmental restrictions. There is little question that these restrictions are resulting in higher capital and operating costs, but it is wrong to assume that the increased costs of producing electricity are due entirely to environmental concerns. Fuel and labor costs, capital expenditures, and interest expenses are all increasing independently of pollution abatement expenditures. The day of cheap electric power has passed. The problem is to separate the environmental costs from the increased expenditures from other sources.

In any case, it is important that the increased costs of producing electricity which are due to pollution abatement be passed forward to the consumers of electric power since subsidized power results in overuse of the product.

Appreciation is given to the Bureau of Business and Economic Research of Arizona State University for granting permission to reprint, in revised form, an article I co-authored with Charles R. Britton, "Standards and Effluent Charges for Pollution Abatement," which appeared in the November 1972 issue of *Arizona Business*.

I express thanks to Michael Butler of Florence State University, Raymond Ziegler of Florida Southern College, Charles Britton of the University of Arkansas, Marvin Ray of Southern Illinois University at Edwardsville, Barry Morris of East Central State College at Ada, Oklahoma, and George P. Smith of the Environmental Protection Agency, for the helpful comments and encouragement in this project. I am particularly indebted to L.D. Belzung, Miles Sonstegaard, and W.A. Halbrook, all of the University of Arkansas, for their criticisms and reviews of the manuscript. Finally, I am grateful to my parents, Robert and Marcia Scott, who made it all possible.

David L. Scott

May 1973
Lakeland, Florida

Foreword

Environmental pollution, a problem that plagues all industrialized nations in varying degrees of severity, has been described as a curse created by today's modern way of life. That is partly true. But, man-made contaminants are as old as civilization—starting with fouled air from early man's fires, or later soot-filled European skies, black from the burning coal and peat as fuel, or, still evident today, widespread nighttime haze from fields burned by Middle East farmers to spread a fertilizing ash.

It is, however, the modern life-style found in industrialized nations that has greatly increased the scope and aggravated the intensity of the problem. The 1970s branded the United States as the world's worst offender. Less developed nations do not have our pollution; but they also do not have our two-car families, our myriad manufactured products, and our array of electricity-consuming appliances—refrigerators and home freezers, clothes dryers and dishwashers, toasters and ranges, vacuum cleaners and other motor-driven devices found all around the home.

As the United States became more affluent, and its expanding population demanded an ever-growing array of products and services, environmental pollution increased gradually until it reached today's disturbing proportions. Only as contamination touched on the lives of each individual did the American citizenry become alarmed. Then, the societal costs that were accumulating all along came to be recognized as obvious costs that the layman could understand.

Transportation and manufacturing are still the greatest offenders; but the electric power industry is not far behind—and it has the dubious distinction of being able to play catch-the-leader. Dr. Scott presents here a thorough economic analysis of this problem.

Someone once noted that Americans crave electricity—not like candy, for what it is; but, like money, for what it does! While total energy consumption in the United States has tended to double every 25 years or so, the rate of electric power consumption has climbed at a far more rapid rate—tending to double each decade. Hydroelectric power causes the least environmental damage; but its potential is very limited. The rapid rise in fossil-fueled electric power generation has made it the worst offender, and—mostly as a result of obstructions by extreme environmentalists—the unfortunate delays in this nation's obtaining its needed electric energy from nuclear-powered plants have served only to compound the problem.

All is not hopeless; but, as this economic study indicates, solutions that will satisfy the American people will be costly and unpopular with some interests.

Professor Scott's timely work is a most welcomed addition to the literature on environmental pollution problems. He adds valuable information, analysis, and insight with his in-depth study. Professor Scott judiciously weighs the plausible alternatives and offers some valid and logical recommendations for both economist and officials who are directly involved in the nation's policy-making process. His findings merit serious consideration by those members of our society who are in a position to "do something about it."

L.D. Belzung
Professor of Economics
University of Arkansas

Pollution in the Electric Power Industry

1 Introduction

The production of electric power is one of the major sources of environmental pollution in industrially-developed countries today. Although the electric power industry is taking steps to reduce the amount of unwanted residuals at plants now in operation and those planned for the future, growth in demand for electricity will bring pressure for an even greater reduction of pollutants. The production and consumption of electrical power has been growing, and is expected to continue to grow, at a compound rate of approximately 7 percent per year. The environmental problems involved will be enormous if this type of increase actually occurs.

By far the largest percentage of electricity currently is generated by fossil-fuel-burning steam plants. The main residuals produced by these plants contribute to pollution of the air through the emission of particulate matter, sulfur oxides, nitrogen oxides, carbon oxides, and hydrocarbons. More recent concern has cèntered on thermal discharges and unsightly plants and transmission lines.

Because of the rapid growth anticipated in electrical output, and the shortages and expected price increases of conventional fuels, primary emphasis has been placed upon nuclear-fueled plants to satisfy future electrical needs. Although initial growth in the operation of these plants has been somewhat slower than expected, nuclear power will assume increasing importance in the years ahead. This, in turn, will lead to a new set of environmental problems. Nuclear operation involves the release of radioactivity and presently discharges even more waste heat per kilowatt-hour of electricity produced than do modern conventional power plants. In addition, nuclear plants must dispose of radioactive wastes. Development is underway now on a liquid-metal fast breeder reactor which, it is estimated, should become operational during the early or middle 1980s. While this type of reactor should be more efficient in producing electricity—and thereby release less thermal waste per kilowatt of electrical power generated—other environmental problems will remain.

This book will examine present and future environmental concerns associated with the generation of electrical power. Cost considerations are included for both fossil-fuel and nuclear plants, along with possible technological improvements that may be forthcoming in generation, distribution, and transmission. Expenses in the areas of capital construction, operation, and pollution control equipment are investigated to determine the effect of pollution control on the cost of electrical power.

Economics will play a major role in determining the type of equipment used to abate undesirable effluents from electric power stations, and how soon the abatement equipment is introduced. Although it is technically feasible to reduce most of the pollutants emitted by fossil-fuel and nuclear-fueled plants to an extremely low level, in many cases it may not be economically desirable to do so.

As with any additional expenses, the problem of who will pay for the pollution abatement emerges. The nature of governmental regulation over the industry is the determining factor in deciding who will absorb the increased costs.

While a wealth of information has been published concerning the environmental consequences of electric power production, most of the more descriptive work has been recorded at congressional hearings and is of limited availability to the general public.

Probably the best source of information for an overview of nearly all facets of the subject is the hearings before the Joint Committee on Atomic Energy in 1969 and 1970.[1] These hearings deal not only with the environmental aspects of electricity generation, but also include an overview of the industry. The testimony includes information on nuclear and fossil-fuel production from almost all interested groups. In addition, the Joint Committee published background material on the environment and electric power prior to the hearings.[2]

Other congressional hearings, dealing with specific aspects of the environment and electric power, also are available. Hearings in 1971 before the House Subcommittee on Communications and Power consider a number of bills relating to power plant siting and environmental protection.[3] The location of new plants has become increasingly important, and proper site selection can help to minimize undesirable effects on the environment and human health. For example, the selection of a location with sufficient waterflow can reduce the need for expensive water-cooling facilities and yet reduce the harm to aquatic life. An earlier site-selection study by the Office of Science and Technology contains much of the same information and is organized for easier reference.[4]

Hearings in 1968 before the Senate Subcommittee on Air and Water Pollution received testimony on thermal pollution.[5] This source is especially valuable in providing information on the harmful effects of waste heat discharges and includes an extensive study by C.B. Wurtz and C.E. Renn on aquatic life and water temperatures.[6] Also, a statement by Tor Kolflat provides an excellent summary of thermal pollution control.[7]

In 1970, hearings before this same subcommittee investigating air pollution provide information on pollutants from fossil-fuel plants—although much of the testimony also concerns emissions from mobile sources.[8] Specific information on pollutants from coal-fired plants can be found in the Senate hearings on the large complex of power stations in Arizona and New Mexico.[9] These hearings provide an insight into how Southern California is able to obtain needed electrical power, yet let others suffer emissions produced by generation of the power outside the state. This method of siting has occurred as power companies are forced to locate new plants in sparsely populated areas, where the companies receive the least criticism from those affected by pollutants. Technological improvements in transmission lines and a desire to locate fossil-fuel plants near abundant sources of coal have speeded this trend.

The Federal Power Commission's *1970 National Power Survey*[10] was begun after the completion of its 1964 survey. The new study is composed of four parts. Part I contains a summary of the six regional reports published in Parts II and III, and a report on the environment. Part IV is composed of committee reports on transmission, distribution, generation, and load forecasting. While the overall report contains much useful information, the reader should keep in mind that nearly all of the members of the various committees submitting the reports were representatives of the power industry. Although Part I is represented as the independent thinking of the Federal Power Commission, it largely reflects the views found elsewhere in the study. The independent Task Force on Environment does include a number of members outside the industry; however, its report comprises a very small portion of the study.

An intergovernmental agency report, *Electric Power and the Environment*, provides a good summary of the institutional setting of the industry.[11] The report attempts to resolve the conflict between environmental protection and the nation's growing power demands, and a number of the study's recommendations were included in President Nixon's environmental program.[12] Hearings before the Senate Subcommittee on Intergovernmental Relations are available on this same topic.[13]

A great variety of literature is available on economics and pollution. Some of the earlier writers in the field—including Allen Kneese, James Buchanan, S.V. Ciracy-Wantrup, and Kenneth Boulding—originated a number of concepts which have not been significantly improved upon by more recent authors. The explanation for this lack of new ideas in economics appears to be that the field is waiting for the supporting disciplines in the physical sciences to "catch up." Economists have developed a framework for weighing environmental and material needs; yet, the costs of damages—let alone complete knowledge of the damages—still are largely unknown. One of the best examples of this is a lack of understanding of how low doses of radioactivity affect humans over long periods of time. Until more concrete evidence is available on this subject, attempts to weigh the benefits and costs of atomic power involve a great deal of subjective judgment.

The use of economics to examine pollution control is explained in Chapter 2. Chapter 3 presents an overview of the electric power industry, including growth trends, financial data, and the methods of power generation. Types of pollutants emitted by electric power stations and their effects on the environment are discussed in Chapter 4. Chapter 5 investigates abatement equipment for controlling residuals and the effect on electricity price of probable control measures. Chapter 6 presents conclusions, recommendations, and a summary.

2 The Economics of Pollution Control

> Clean air, clean water, open spaces—these should once again be the birthright of every American. If we act now—they can be. We still think of air as free. But clean air is not, and neither is clean water. The price tag on pollution control is high. Through years of past carelessness we incurred a debt to nature and now that debt is being called.
>
> President Richard Nixon
> *State of the Union Address*
> January 22, 1970

The increasing concern over pollution[1] of our environment can be described as little short of dramatic. Only a few years ago, environmental considerations were given minor—if any—priority in most decision-making processes. Although pollution is nothing new, much of the recent reaction results from an increasing magnitude of the problem. Americans now have more cars, more factories, and more people making use of an earth environment of fixed size. This environment's assimilative capacity to absorb many of the pollutants is being strained and surpassed.

While some environmentalists attribute the major share of blame for America's plight to population increases, and others criticize faulty and misdirected technology,[2] few can doubt that the problem is real and will surely worsen unless positive corrective action is taken. It is the form and size of this action that occupies much current debate on environmental pollution. Proposals have ranged from one extreme of doing nothing to the other extreme of zero effluent discharge. This has been accurately described as the "fallacy of all or nothing."[3] Economics, integrated into an interdisciplinary framework, can help shed light on some pragmatic solutions to this pressing problem.

Economists generally explain the pollution problem as originating from a discrepancy between social and private costs of production. Since product prices take into account the cost of resources to the manufacturer, but not necessarily the full costs to society, people may be receiving too many or too few of some goods and services and they may be produced by socially undesirable methods. For example, one can consider a municipal sewage plant that only partially treats effluents before releasing them into a lake that is used for fishing. If the amount of untreated sewage is so great as to reduce the fishing productivity of the lake, society will pay a lower cost for faulty municipal sewage treatment and a higher price for fish than otherwise would be the case. This inequality between social and private costs is termed a "spillover" or "externality."[4] In the case just described, the sewage plant results in an external diseconomy, since its operation means increased costs for other sectors of society. The plant is using a common property resource to the detriment of other users and/or potential users. In addition to a reduction in fish catch, the sewage may make the lake unfit for swimming, boating, and using as a water supply source. Proper sewage treatment to give the lake additional uses may be relatively inexpensive, but management has little incentive to implement pollution reduction measures since there is no direct benefit from its doing so.

The problem with a collective good[5] is that its use by an individual does not entail the full (or possibly any) payment by the user, but does affect others who have an equal right to its amenities. Equally important, the full benefit of any improvements will not accrue to those undertaking the improvements. These are the external costs and benefits of the externality problem. If the charge for this resource is minimal or nonexistent, the user can be expected to use the common-property resource extensively. Such has been the case with air, lakes, rivers, and many land areas as they acted as pollutant receptors. In many cases, a single collective good may be used as two completely separate goods by two different interests. The fishermen want the lake as a clean water resource to provide fish, while the municipal sewage plant desires the lake as an effluent receptor regardless of the water's condition.

Depending upon how laws are structured (whether a community has collective rights to common property resources or whether resource uses are without restrictions), payoffs may take place between the polluters and other users. In the latter type of law without restrictions, fishermen might find it worthwhile to pay the sewage facility an amount sufficient to reduce residuals to an acceptable level. The smaller the cost of adequate pollution control, relative to the value of other uses, the more likely this payoff is to take place. The inhibiting factor here is the free-rider problem; alternative users might wait for others in the group to incur the payoff expense, and so it might be delayed or not take place at all. In the former example of a community's right to a common property resource, the payoff is more likely when the cost of adequate pollution abatement equipment is higher relative to the value of alternative uses. The free-rider problem does not exist in this case, and the actual payoff is more likely. Most laws today favor the polluter, rather than alternative users, and so the "community right" example is not prevalent.

The payoff solution has its roots in the Pareto criterion of welfare maximization.[6] Under this criterion, a change which makes one individual better off but leaves no individual worse off increases the collective welfare. Described in this narrow manner, it is of little help in one's solving environmental problems. However, if the criterion is interpreted more broadly to allow compensation from gainers to losers, it becomes more useful as a tool in pragmatic problem solving. The compensation is in the form of the payoff described. If the cost of pollution control is relatively high, the compensation to alternative users presumably must be of such an amount that they are indifferent between use of the resource and the payoff. Hence, the polluter has improved his welfare because he is not required to meet the high expenses of pollution control, and those who suffer from the pollution are fully compensated. Alternatively, and possibly more likely, the polluter might decide it would be less expensive to undertake partial pollution abatement in order to make the water fit for uses not requiring complete cleanup, and pay the remaining receptors who require water of higher quality. In the case of the sewage disposal plant, the control measures might permit boating and certain types of fishing, but not swimming or using the lake for a supply of water.

Partial control is likely because of the cost structure of abatement facilities. In general, the cost per unit of pollutants removed at low levels of control is much less than the cost per unit removed at high control levels. For example, Douglas Caddy, using data supplied by the Council on Environmental Quality and the Environmental Protection Agency, estimated the costs of cleaning the environment to be as shown in Table 2-1.

Although the estimates in Table 2-1 are undoubtedly "rough," the point seems clear that, as higher levels of abatement are approached, the job of control becomes increasingly more costly and complex. In many cases, the extremely high control levels might be neither necessary nor particularly desirable.

Using this type of analysis, economists generally contend that pollution control measures should be a function of the costs and benefits in each particular case being considered, since the resources devoted to very high purification levels might best be used elsewhere for greater benefits to society. Many economists, therefore, sided with Nixon Administration officials in denouncing the pollution control bill introduced into the United States Senate by Senator Edmund Muskie, requiring zero effluent discharge into waterways by 1985.[7]

The solution to the externality problem is to introduce a market-type apparatus into any area where the market is not already operating. In other

Table 2-1
Total National Costs of Successive Levels of Pollutant Removal (Billions of 1971 Dollars)

Level of Removal	Total Cost	Cost per Incremental Percentage Point of Removal
85%–90%	$ 61.0	$ 0.7
95%–99%	119.0	6.0
100%	317.0	66.0

Source: Douglas Caddy, "Effluent Society?" BARRON's, LII, No. 5 (January 31, 1972), p. 5.

words, externalities should be internalized into the polluter's cost structure. A number of alternatives have been suggested to cope with environmental problems, but most of these fall into two categories—taxation (or effluent charges) and standards.

Effluent Control Through Standards

The setting of standards has been the most common approach used to reduce emission of undesirable residuals. Standards have been applied to cover such pollutants as hydrocarbons, particulate matter, carbon dioxide, sulfur oxides, waste heat, and various emissions into water. They also have been set by the Atomic Energy Commission, in conjunction with the Federal Radiation Council, in the area of radioactive release.

Theoretically, quality standards should be set in such a manner that the cost of pollution damage plus the cost of pollution treatment is minimized. This would be the optimal standard. No fine is levied against effluents which are emitted once the standard is met. Standards more strict than this will entail a pollution control cost greater than the benefit derived from the expenditure. This type of solution is illustrated in Figure 2-1 and in the hypothetical cost figures of Table 2-2, from which the graph is derived.

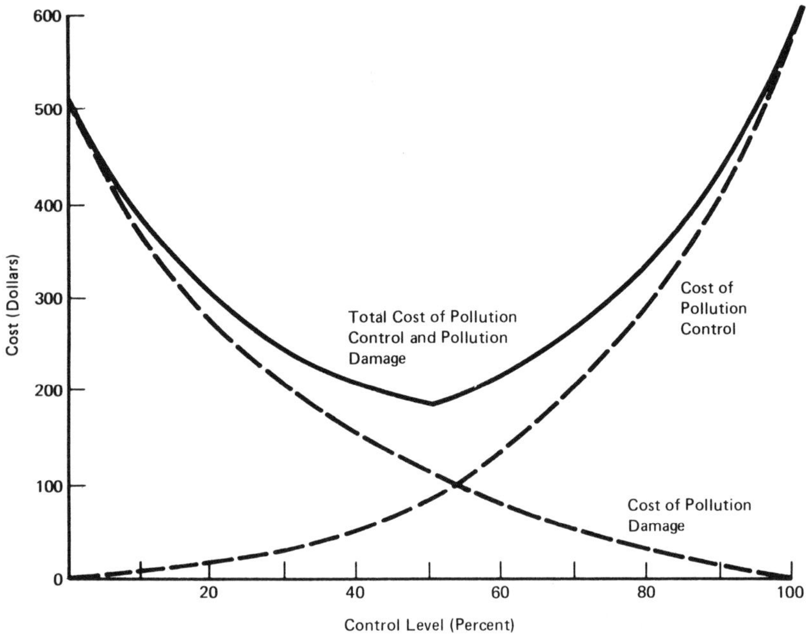

Figure 2-1. Hypothetical Costs of Various Levels of Pollution Control.

Table 2-2
Hypothetical Costs of Various Levels of Pollution Control

Control Level (Percent)	Cost of Pollution Damage	Cost of Pollution Control	Total Cost
0	$510	$ 0	$510
10	385	5	390
20	285	15	300
30	205	30	235
40	155	55	210
50	105	90	195
60	70	140	210
70	45	210	255
80	25	300	325
90	10	410	420
100	0	600	600

Source: Adapted from Azriel Teller, "Economic Rationality and Reality," DAEDALUS, JOURNAL OF THE AMERICAN ACADEMY OF ARTS AND SCIENCES, XCIV, No. 4 (Fall, 1967), p. 1086.

Figure 2-1 shows that in this particular case, the optimal control level standard is 50 percent of total emissions, since the total cost of control plus damage is at a minimum of $195. Although one-half of the unwanted residuals still will be released after the standard is enacted, further emissions reduction will cost more than the benefits derived from such added reduction. The estimated cost figures shown in Table 2-1 would make the pollution-control cost schedule much more steeply sloped at high levels of control and, thus, make high control levels less desirable than is shown in Figure 2-1 and Table 2-2.

As more efficient methods of pollution abatement are developed, the cost-of-control schedule would fall, dictating higher control standards. New data concerning damage from the pollutants would shift the cost-of-pollution-damage schedule to the left or right, depending upon the information obtained. This also would call for new standard setting. It should be noted that this analysis does not preclude an increase in pollution over time. As population increases and the economy expands (assuming static technology), both schedules will rise as the cost of pollution damage and the cost of pollution control increase at given control levels. Depending upon the relative increases of each schedule, the optimal standard may be higher, lower, or the same as before the change. In any case, the cost of each will be greater and the absolute amount of effluents discharged and controlled will increase. This will be so even if the percentage of emitted residuals remains the same.

A similar method of analyzing standards is through a comparison of the marginal benefit and marginal cost of pollution control. Figure 2-2 illustrates that, as the standard or control level is increased, the marginal social cost of pollution treatment will approach, equal, and then exceed the marginal social benefit of treatment. The optimal level of abatement in this case is at a control level of *X* percent. At standards more severe than *X* percent, the benefit is exceeded by treatment cost, and society could improve its welfare by devoting to other areas a portion of the resources used for pollution control. The optimal control level in Figures 2-1 and 2-2 would be identical for the same standard.

A number of complications surface if standards are a primary pollution control mechanism. The first and foremost problem is the difficulty in quantifying environmental damage due to pollutants. A few rough estimates have been made, but they are not very accurate. For example, probably the most widely quoted estimate of annual damage from air pollutants from all sources is $65 per capita; however, the method of arriving at such a figure is highly suspect, and many believe that the actual damage is much greater.[8] Other estimates put the annual cost of air pollution damage to livestock and agricultural crops alone at $500 million or more.[9] An even more difficult problem arises when much of the damage is impossible to quantify. What value can be placed on human life? What is the environmental damage due to putting high voltage electrical transmission lines above ground rather than below ground? Scientists do not even agree on whether continuous low-dose radioactivity is harmful to humans, let alone the dollar damage due to it. Without knowledge of the physical effects of pollutants, one cannot begin an accounting of their actual dollar costs.

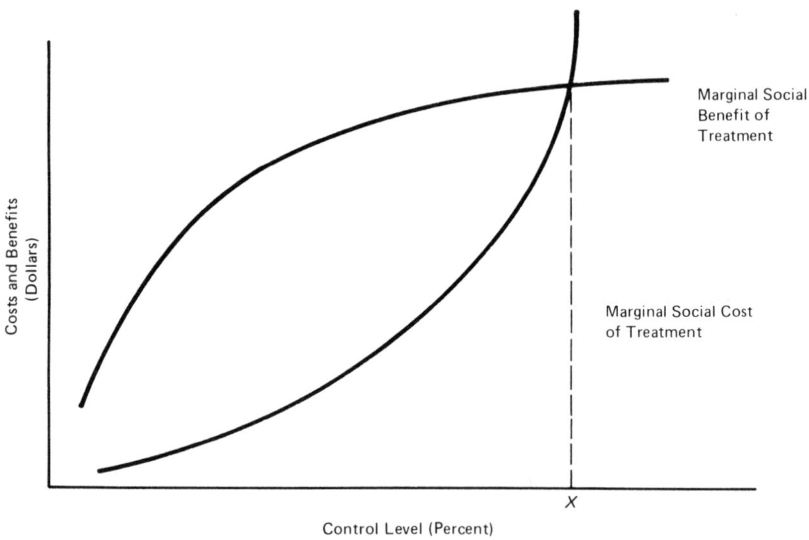

Figure 2-2. Marginal Social Benefit and Marginal Social Cost of Pollution Control.

Another disadvantage of there being a standard is that, once it is met, there is no incentive to reducc cmissions below the prescribed control level. Although the development of better pollution control devices might take place outside the emitting industry in order to stimulate sales to polluters through a forced increase in standards (improved control techniques would bring about higher control level requirements), little research and development can be expected from the offending industry. Unfortunately, in many cases, it is the industry releasing the pollutants that is best equipped to develop adequate methods to combat the problem.

Another difficulty with a standard as a method of pollution control is its inflexibility in application. Once the standard is met, each change in requirements might entail the replacement (with new equipment) of most of the pollution-abatement equipment already installed. If the penalty for not meeting the new standard is a large fine or the shutdown of operations, the transition period could prove difficult for the companies involved. If a period of grace is allowed for installation of the new equipment, the polluters have no incentive to beat the deadline.

The most important economic drawback of a standard is the inefficient allocation of resources that it entails. Since firms have different marginal cost schedules for pollution abatement, a given control level for all firms causes much greater costs for some firms than for others. Abstracting from administration costs, the same net result of pollution control could be obtained more cheaply if various levels of control were set for each company on an individual basis. Likewise, a greater amount of pollution abatement could take place for the same amount of money. An example may serve to clarify this consideration. Suppose a city or a region decides that the emission of particulate matter should be reduced by 80 percent and, therefore, it applies an identical standard to each polluter within its jurisdiction. Some companies presumably will be able to meet the reduction schedule at nominal cost because of the nature of their production functions, while other companies will find it exceedingly expensive to meet the same standard for the same reason. The same overall reduction could be induced at less cost by requiring greater reductions by the companies facing more elastic marginal cost schedules for pollution control and smaller reductions by the companies with high marginal control costs. Hence, the companies facing the most severe control-level restrictions should be those that could do so with the least cost. Under a standard applied to all companies, the inequality of individual cleanup costs does not enter into the decision-making process. Figure 2-3 illustrates this point. In Case *A*, the same 80 percent control level is set for Company *F* and Company *G* who have marginal cost schedules of MC_f and MC_g, respectively. Provided these are the only two companies affected under the standard, the total cost of pollution abatement is the summation of the areas to the left of the 80 percent level under each of the curves. Case *B*, however, shows that the same 80 percent overall reduction can be met by a requirement that Company *G* meet a more strict 96 percent level and Company *F* a more lenient

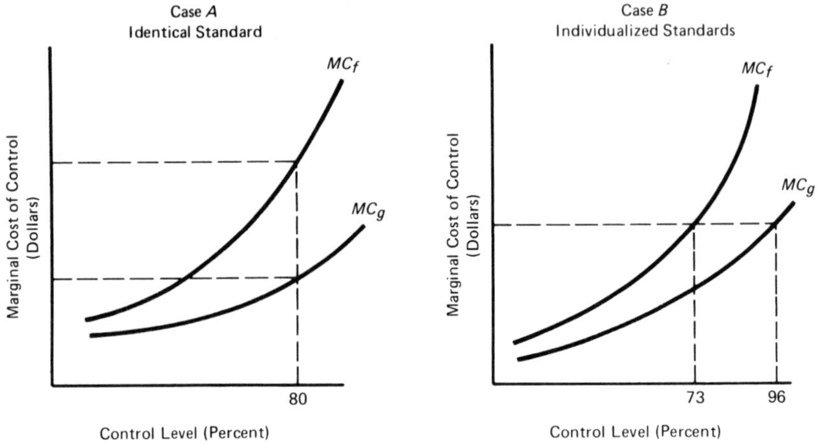

Figure 2-3. Relative Cost Comparison Between a Single Standard and Standards Set on an Individual Basis.

73 percent level. In practice, the weights of the two companies would have to be taken into account in the setting of individual standards. The total cost of control by this latter method is the area under MC_f to the left of 73 percent plus the area under MC_g to the left of 96 percent. The total cost of the individualized method of standard setting is less than the total cost of the identical standard setting, although the same net result of pollution abatement is achieved.

Another variable more easily accommodated by individualized standards is location. Most pollutants are considered more harmful when released in densely populated areas. Under the more flexible individualized standards, lower control levels could be required of polluters in sparsely populated areas. With an overall reduction goal set at a certain level within a region, polluters near population centers could be forced to meet rigid standards. It should be noted that locational considerations and cost considerations will not necessarily coincide.

Unfortunately, the individualized standard setting would prove extremely difficult in practice because the regulating authority requires knowledge of each polluter's cost functions in order to develop a simultaneous relationship for distribution of the quotas. Any change in the cost function of one firm would automatically affect the other firms and bring about a new least-cost solution.

Pollution Abatement Through Effluent Charges

Most economists favor taxation or the imposition of effluent charges as a method of internalizing external costs. The chief criticism of this alternative has come from environmentalists who say that the effluent charge allows offending companies to buy the "right to pollute." Although it is correct that companies can buy the "right to pollute" under this system, the statement, itself, is not a valid criticism of taxation because it enables society to achieve the same results as a standard with less administrative difficulties.

The chief advantage of an effluent charge is that it provides for a more efficient allocation of resources than does a uniform standard. Kneese and Bower have estimated that, for water quality control in the Delaware Estuary, an effluent charge system would produce a given level of water quality for about one-half the cost of a uniform treatment method.[10] The efficiency would be the same as under the individualized standard method; however, taxing does not require a regulating authority to apportion quotas on the basis of each company's marginal cost-of-control schedule. Rather, the simultaneous least-cost solution will be obtained as each polluter decides upon the proper control level to minimize the total cost of taxes plus pollution-control costs. Those companies which experience a relatively high cost per unit of emission-reduction will prefer to undertake low control levels and pay the tax, while companies which have only nominal reduction costs will eliminate a high percentage of their residuals. The control levels decided upon will be a direct function of the size of the tax, and, for more complete total emission control, the regulating authority needs only to increase the size of the effluent charge. Each company will experience a least-cost solution when the cost of reducing one more unit of the effluent is equal to the tax per unit of residual released. In other words, the marginal cost of control will equal the tax at the optimal level of abatement. This equality for three hypothetical firms is illustrated in Figure 2-4. Firm *A* has a relatively high marginal-cost-of-control schedule and finds its best solution is only a 30 percent reduction in emissions, while Firm *C*—with low control costs—has an optimal

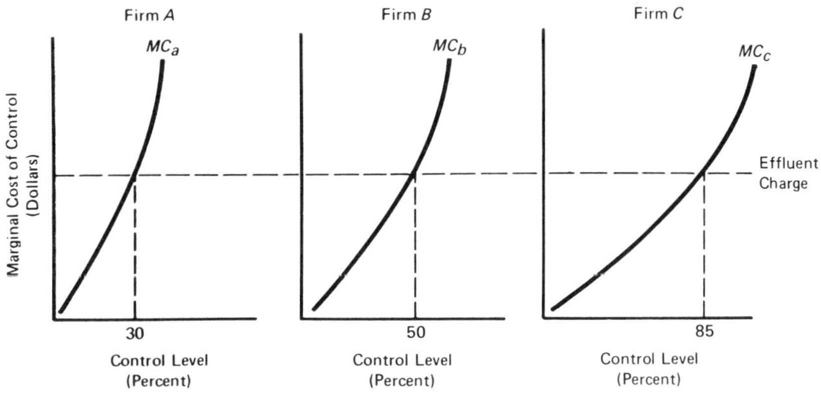

Figure 2-4. Emission Control Through Effluent Charges.

emission reduction of 85 percent. Firm *B* possesses a marginal cost schedule between the other two firms and reduces its pollutants by 50 percent. The same aggregate control level cannot be obtained more efficiently at any other combination of control levels by the three firms. Should greater control of residuals be desirable, the pollution authority would raise the tax rate on emissions. Each of the three firms then would move upward on their respective marginal cost curves until the new control level is reached. At this point the marginal cost of control will equal the effluent charge.

There are other advantages inherent in a system of taxation for the control of pollution. Since the regulating authority does not need each firm's cost data and does not need to set individualized quotas, the cost of administration should be greatly reduced. Each time that the authority decides to alter total emissions, a cost saving will accrue under the effluent-charge type of system, since there will be no need to review each company's data in order to apportion new quotas. Also, relative cost changes between companies would be less important under this system. In addition, the revenues generated by the tax could be used for administrative expenses and collective research and development or environmental improvement.

There are two primary difficulties involved in the use of effluent charges. The first, as with a standard, is the near impossibility of obtaining a dollar accounting of damages incurred because of the emission of residuals. There are a number of extreme areas, including death and aesthetic costs, where no quantitative method of measurement is available. Presumably some arbitrary values must be assigned to these variables, with the recognition that they are highly subjective. Some of the other damages, such as health, although partially quantifiable, still must be considered in the realm of subjectivity. Possibly, the known damage costs could be accounted for and a variable percentage could be agreed upon for intangible costs, to be added on for an approximation of the total cost of pollution damage.

The second difficulty is in determining at what level the effluent charge should be set. Without reasonably accurate estimates of pollution control costs and pollution damage costs, the effluent tax rate must be set in a subjective manner. At times, taxes or standards might be invoked against pollutants for which no known remedial technology is available. This puts the polluters in the position of paying the tax or reducing production, because no other choices are available. A current example closely resembling this is the control on sulfur oxides. In addition, controls in the form of standards or taxes have been used, even though the damaging effects of the pollutants are unknown. Despite this drawback, however, the flexibility of taxation allows the tax level to be raised or lowered to achieve more closely the desired control level.

An approach similar to the taxation on effluents is the subsidization of pollution abatement measures. With subsidies, however, the question of who pays for the control is in no way related to the question of who causes the pollution.

Examples of subsidies are low-interest loans, tax reductions, or direct payments to firms undertaking steps to reduce emissions. The problem with subsidies is that, unless the polluter is fully reimbursed for the cost of control, there will be little or no incentive for him to engage in the reduction, since it would be an added expense that would produce no additional revenues. A more likely use of subsidies is in conjunction with emission standards. This is occurring to some extent now, since municipalities have been allowed to issue industrial revenue bonds for abatement equipment. This effectively lowers the cost of the equipment through lower interest rates, due to the tax-free status of the interest payments for federal income-tax reporting purposes. Companies use this financing only because they are required to meet emission standards.

The Burden of Pollution-Control Costs

One of the basic problems in pollution abatement is the determination of who will pay the control costs. Many environmentalists feel that the total cost of control should be absorbed by the entity releasing the residuals, since there would not be any pollution problem if it wasn't for the polluters. Although it undoubtedly is correct that polluters cause pollution, the conclusions drawn from this concept are somewhat suspect. In reality, this seems to be an example of the more general response that "someone else should pay for the cost of environmental cleanup." This attitude probably is the reason why less public pressure has been thrust upon municipalities for pollution control than on private concerns, since there is little doubt who will be forced to bear the cost in the case of municipalities. However, there is some evidence that consumers are willing to pay at least part of the cost for obtaining a cleaner environment. One poll showed that 70 percent of the people surveyed would be willing to pay 25 cents per month on their utility bill for a complete cleanup of emissions from electric power companies.[11] Although this sum is not large, at least it demonstrates some interest by consumers to pay a portion of the cost.

One of the fundamental economic premises of externalities and pollution is that the producer with external diseconomies is not paying the full costs of production. This low cost, in turn, usually results in an artificially low price to consumers, and means that they will be purchasing too many units of the finished product (assuming an elasticity of demand greater than zero). Only when the producer pays the full social cost of production will the price and quantity purchased reflect the externalities.

Pollution Control in Perfectly Competitive Industries

If a tax is used to stimulate abatement, the portion of the tax that is passed forward to consumers in the form of increased prices depends upon the type of demand curve faced by the firm being taxed. In the case of a firm under perfect competition and facing a horizontal demand curve, all of the tax eventually will be paid by consumers. Figure 2-5 illustrates this concept. Assume that the firm is originally in long-run equilibrium, producing quantity $0q_1$, and that the industry is producing quantity $0Q_1$ at price P_1. If a per-unit charge on effluents is imposed by a regulating authority, the firm will have new cost curves as MC_1 shifts upward to MC_2, and AC_1 shifts upward to AC_2. The upward movement of each curve will be by an amount equal to the tax, and the short run equilibrium output for the firm will be $0q_2$, where MC_2 equals marginal revenue or price. However, under these conditions, the company will be suffering an economic loss, on a per-unit basis, equal to the difference between AC_2 and price at output $0q_2$. In this short run equilibrium, the firm is incurring the entire cost of the tax. In the long run, as firms begin withdrawing from the industry, the industry supply curve shifts to the left, forcing product price higher. Firms will stop leaving the industry when a new price that equals the low point on AC_2 is established. At this price, the firm is producing at its old output $0q_2$, but the industry output is now $0Q_2$ because of fewer firms. Hence, under conditions of perfect competition, the firm absorbs all of the tax in the short run, but the consumers pay the tax in the form of higher prices in the long run. Purchases of the industry's product will be reduced owing to the higher price and elasticity of demand, and release of the pollutants associated with the product will decline because of lower output. Workers in the industry will bear a portion of the cost in that fewer employees will be required for the lower production.

If a standard is set and the penalty for a firm's not meeting the standard is a

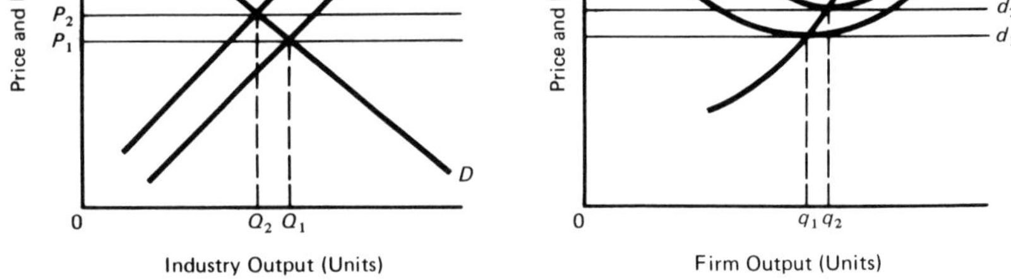

Figure 2-5. Pollution Regulation by Taxation Under Perfect Competition.

fine, the result will be an upward shift of the average cost curve, with no change of the marginal cost curve. This type of regulation is illustrated in Figure 2-6. Assume the firm has costs associated with AC_1 and MC_1 before the standard is imposed. The firm's output is $0q_1$ and industry output is $0Q_1$. Once the standard is invoked and the firm pays the fine rather than install pollution control equipment, AC_1 shifts upward to AC_2. Because of the nature of the lump-sum fine, it is a fixed cost for the firm, and average cost rises more at low outputs than at high outputs. As with a per-unit tax, the firm will lose money in the short run, but as competitors begin leaving the industry, S_1 shifts upward to S_2 until the new long-run equilibrium price is equal to the low point on AC_2—thereby wiping out the firm's loss. At price P_2, industry output is $0Q_2$ and firm output is $0q_2$. Again, the firms pay the cost of the standard in the short run, but the whole cost is passed on to consumers in the long run. Employment in the industry declines owing to a smaller output.

The foregoing analysis assumes that firms decide to pay the tax or fine, rather than install pollution abatement equipment. But, what if the equipment is actually installed? The determination of cost changes because of equipment installation is not known unless the equipment cost structure, itself, is known. If the equipment entails only a capital expenditure and no operational costs (per-unit of output), there is a fixed charge that shifts average cost up and does not change marginal cost. If, on the other hand, the opposite is true, and operational costs vary proportionately with output, the equipment has the same effect as a per-unit tax, resulting in an upward shift of both average cost and marginal cost.

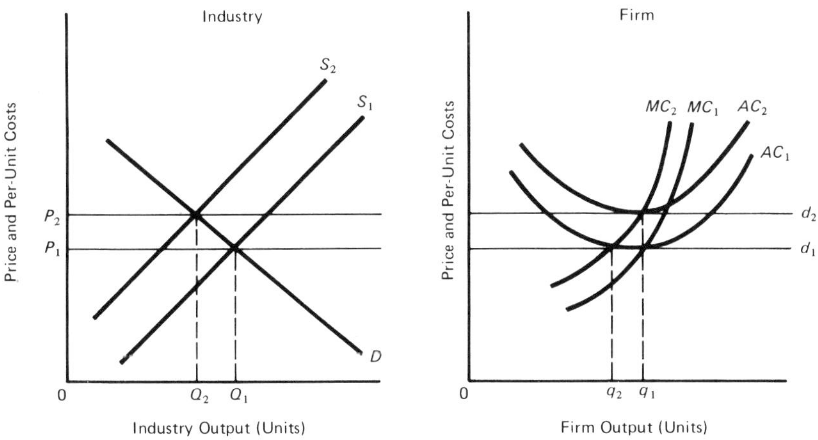

Figure 2-6. Pollution Regulation by Standards Under Perfect Competition.

Hence, with conditions of perfect competition, all costs will be passed on to the consumer in the long run, regardless of whether the firm pays a tax, a fine, or actually installs the pollution reduction equipment. In addition, industry output and employment will be reduced.

Pollution Control in Industries Not Characterized by Perfect Competition

Under industry structures other than perfect competition, the determination of who will pay the costs of pollution control is not so clear-cut. These firms do not face a horizontal demand curve and so part of the costs might be passed forward to consumers or backward to labor (lower wages), to stockholders (lower dividends), or to society (subsidies and tax write-offs).

First assume a per-unit tax is used to promote the use of abatement facilities. Figure 2-7 shows a firm in equilibrium producing $0q_1$ units at price $0P_1$. It faces AC_1 and MC_1 before imposition of the tax. Once the per-unit effluent charge is levied against the firm, MC_1 and AC_1 shift upward to MC_2 and AC_2, which is

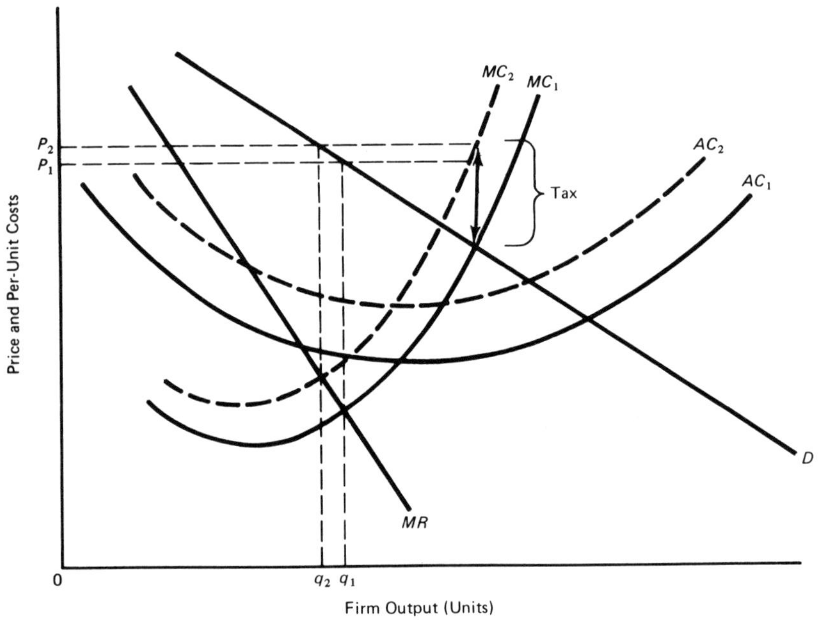

Figure 2-7. Pollution Regulation by Taxation in Industries Not Under Perfect Competition.

the amount of the tax on a per-unit basis. Since MC_2 intersects MR at a different level, the profit maximizing level of output is reduced to $0q_2$ and price rises to $0P_2$. The portion of the charge passed forward to the consumer is the price rise, while the polluter must "bear" the remaining part of the tax. Hence, a per-unit effluent charge in an industry not facing perfectly competitive conditions is borne partially by the consumer and partially by the polluting company. As under perfect competition, output is reduced and industry employment falls.

If a standard is used to entice these firms into using control equipment and if the penalty of not meeting the standard is a lump-sum fine, the result is an upward shift of average cost and no change in output. Figure 2-8 illustrates the use of a standard under less than perfectly competitive conditions. Assume the firm is maximizing profits at $0q_1$ units of output and charging price $0P_1$. Since it faces AC_1, MC_1, and D, its profit level is the vertical difference between price and average cost (AC_1) times output $0q_1$. If a standard is imposed and not met, then AC_1 shifts upward to AC_2 to reflect the fine. Although profit is reduced, the firm will not cut back on production since there has been no change in marginal cost and $MR = MC_1$ at the same output level. Since the fine on a per-unit basis is reduced as output increases, there is a greater difference between

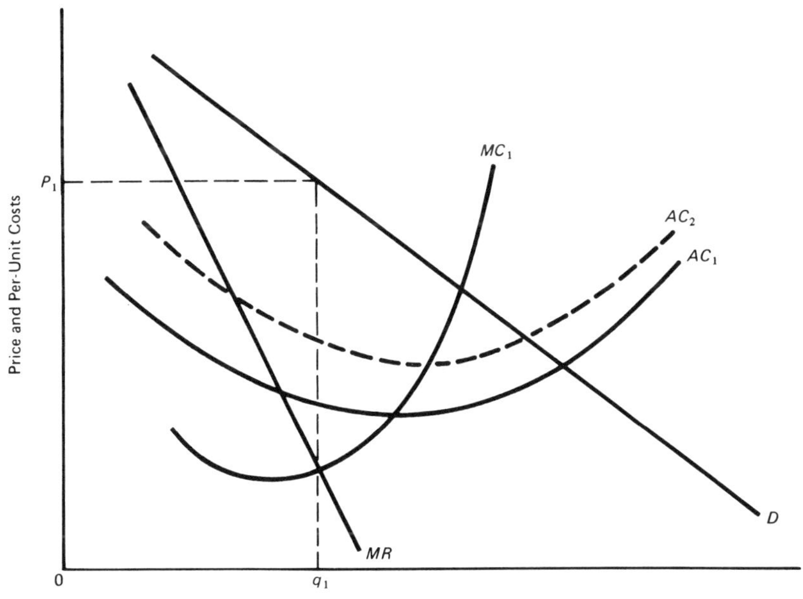

Figure 2-8. Pollution Regulation by Standards in Industries Not Under Perfect Competition.

AC_1 and AC_2 at lower output levels. In this case, the entire fine is borne by the firm (shifted backward) and none of it can be shifted forward to the consumer in higher prices. Unless the fine is so great that it forces the company out of business, no reduction in industry output and employment takes place. Should some companies leave the industry without new ones entering, the consumer will pay higher prices since the industry supply curve will shift leftward. If the penalty for not meeting the standards is a shutting down of operations, the supply curve also may shift leftward.

If the company decides to install abatement equipment—rather than pay the fine or tax—the effect is indeterminable, unless the cost structures of the devices are known. As under perfect competition, a capital charge is a fixed cost that leaves short-run marginal cost unchanged. This lowers the firm's profit, but leaves price and output unchanged, and so the cost is borne only by the polluter. If the equipment requires operational expenses varying with the amount of pollutant reduction, there will be a change in marginal cost and so the company and the consumer each will be forced to absorb a portion of the costs. The proportional share will depend upon the elasticity of the demand curve that the firm faces. If, as is likely, the abatement equipment entails both of the foregoing types of costs, both the consumer and the company will pay.

Administered Pricing

If firms practice administered pricing, as many economists believe, then one can expect consumers to pay not only for the pollution control equipment, but also an additional markup. For example, suppose a company prices its product to earn 10 percent on invested capital after taxes. If the firm installs $100,000 worth of control equipment, it raises the price of its product not only to cover depreciation and operational expense, but also enough to earn some percentage of return on the equipment itself. In this case, the consumer not only is paying for the cost of pollution control, but he also is paying higher prices so that the firm can earn a target return on the control equipment. The assumption of administered pricing is that firms do not attempt to maximize profits.

The Economic Impact of Pollution Control

The impact of the increasing amount of resources devoted to effluent abatement will affect all members of the American society. As shown previously in this chapter, the costs of control will be spread throughout the economy, but it is difficult to determine who will bear the major portion of the burden. This will depend upon the type of controls used, the kinds of industry structures being affected, and the nature of the costs of control equipment. Although the installation of the equipment will decrease the amount of consumer goods and services available to society, this should not be confused with a resulting lower standard of living. For example, an Atomic Energy Commission official recently commented, "Such persons (environmentalists) contend that if we are to preserve a habitable earth, population growth must be stopped and members of society must be willing to expect fewer goods and services—a lower standard of living—as the price of protecting the environment."[12] The proper amount of control will result, not in a lower standard of living, but a higher standard of living, through the replacement of goods and services by cleaner air, cleaner water, and improved health. Also, it will create jobs for those who make and service the new equipment; there is a reallocation of certain resources, but not a reduction in total resource utilization.

3 An Overview of the Electric Power Industry

> On Sunday, February 7, 1971, a power failure in the Consolidated Edison system plunged half a million New York residents into a blackout for several hours. Ironically that Sunday was the beginning of "Electricity Week," proclaimed by Governor Rockefeller in praise of "the increasing contribution the electrical industry is making for the betterment of our lives."
>
> *Home News*
> New Brunswick, N.J.
> February 24, 1971

The above news item typifies both the promise and the problems of the electric power industry in the United States today. Although a social and industrial necessity, the production of electricity is confronted by increasing public concern in all areas of the country. Not only are consumers unhappy, but most shareholders also have shown displeasure owing to a bleak decade in terms of capital appreciation on investment. The great bulk of the problems results from the seemingly insatiable appetite for more electrical energy in nearly all sectors of modern society. Past and projected trends in electric power production show almost a 7 percent annual rate of increase, with a doubling time of approximately one decade. While most industries and investors would envy such a steady and assured growth pattern, this has not been typically the case for the electric power industry. This chapter will provide some insight on this and other aspects of the vital electric power industry.

Increasing use of electricity is a direct function of the rising demand for energy resources in America's industrial society. People need greater amounts of energy resources to travel more, to produce more goods, and to enjoy the "good life." The increasing consumption of primary energy over the past decade is displayed in Table 3-1.

Five primary sources furnish the energy that performs a majority of work in the United States. These are coal, oil, natural gas, radioactive ores, and water power. While electricity is an important form of energy, it is secondary rather than primary since its production requires one of the primary energy sources. Electricity is merely a more convenient form of delivering and consuming energy for many uses. Although some of the primary sources are more important than others in producing electric power, all are used to some degree, and anything that affects electric power production will in turn affect the primary sources of energy.

Table 3-1
Consumption of Energy Resources by Major Consumer Group: 1963-1970 (Trillions of Btus)

	Energy Inputs				
Consumer Group	1963	1965	1967	1969	1970 (Prel.)
Household and Commercial	11,059	11,867	13,054	13,628	14,098
Industrial	16,225	17,734	18,540	20,570	21,129
Transportation	11,964	12,715	14,150	15,950	16,436
Electrical Generation, Utilities[a]	9,663	11,104	13,135	15,433	16,967
Miscellaneous	738	549	277	192	180
Total	49,649	53,969	59,156	65,773	68,810

[a]Represents outputs of hydropower and nuclear power only.

Source: U.S. Department of the Interior, Bureau of Mines, MINERALS YEARBOOK, as cited in U.S. Department of Commerce, Bureau of the Census, STATISTICAL ABSTRACT OF THE UNITED STATES: 1970, p. 506, and 1971, p. 496.

Because of the convenient nature of electricity and the fact that it is cleaner than the primary energy forms at the point of consumption, its use has been growing at an even more rapid rate than energy use in general. Former applications of primary energies are being replaced more and more by electricity in final consumption. Nowhere is this more evident than in the case of residential heating. Many new homes are being constructed to use electrical resistance units in floors, walls, and ceilings, in place of fossil-fuel-burning furnaces. The four major markets for primary energy include industrial-commercial, transportation, electric utilities, and residential. In 1950, the electric utility industry was the smallest of the four major primary energy markets. By 1965, it accounted for nearly 21 percent of the total and moved up to third place. By 1980, utilities are expected to be the largest single user, consuming almost one-third of the total production from primary energy sources.[1] Figure 3-1 illustrates a projection of the increasing importance in converting primary energy to electrical energy.

The production of electrical energy has been doubling approximately every ten years. Table 3-2 illustrates this phenomenal record, along with the type of ownership of the producing units. This rate of growth leads to some disconcerting conclusions, in that the increase in electrical production over a ten-year span must approximate the total output only a decade earlier. If one assumes that the same percentage of capacity is used in production over time, it also means that productive capacity must double each decade. The Federal Power Commission

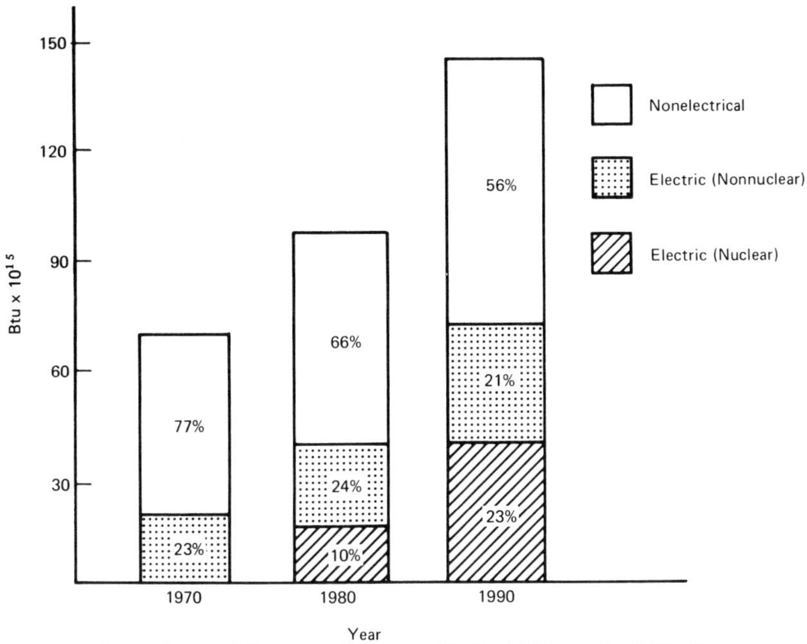

Figure 3-1. Users of Primary Energy: 1970, 1980, and 1990. Source: U.S. Atomic Energy Commision, *The Nuclear Industry* (Washington: U.S. Government Printing Office, 1972), p. 9.

estimates that by 1990 the nation's electric power industry will have to build and finance approximately 900 million kilowatts of new electrical generating capacity.[2] This is nearly three times the capacity available in 1970.

Methods of Producing Electricity

As the electric power industry has expanded, the composition of primary energy used to produce the electricity has changed in character. Table 3-3 shows the sources of energies used from 1940 to 1970 in the production of electricity. As can be observed in Table 3-3, hydropower is declining in relative importance at the expense of oil and natural gas, while coal has remained approximately 50 percent of the total over the past three decades. Coal, oil, and gas are all fossil-fuels and generally are burned to produce heat, which converts water to high pressure steam used to turn turbines and generators. Oil is used in some instances in an internal combustion process to power generators directly rather than indirectly through steam production. In the future, nuclear-steam generating plants will play an increasingly important role. Although a few nuclear plants have been in operation since the middle 1950s, a large number are

Table 3-2
Electric Power Production in the United States by Class of Ownership, 1940 to 1970 (Billions of Kilowatt-Hours)

Year	1940	1945	1950	1955	1960	1965	1970 (Est.)
Industrial Plants	30	49	60	82	88	102	108
Electric Utilities (For Public Use)	142	222	329	547	753	1,055	1,530
Privately Owned	125	181	267	421	579	809	1,183
Publicly Owned	16	42	62	126	175	246	347
Municipal	6	10	15	26	37	50	71
Federal	9	28	40	89	112	145	186
Cooperatives	1	3	6	11	26	51	89
Total	180	271	389	629	842	1,158	1,628

Source: U.S. Federal Power Commission, PRESS RELEASE NO. 16634, as cited in U.S. Department of Commerce, Bureau of the Census, STATISTICAL ABSTRACT OF THE UNITED STATES: 1971, p. 497, and 1970, p. 507.

Table 3-3
Sources of Energy Used in the Generation of Electricity (Percent)

Year	Coal[a]	Oil	Gas	Hydro
1940	54.6	4.4	7.7	33.4
1945	51.7	3.5	8.9	35.9
1950	47.1	10.3	13.5	29.2
1955	55.1	6.8	17.4	20.7
1960	53.6	6.1	21.0	19.3
1965	54.5	6.1	21.0	18.4
1970 (Est.)	46.4	13.2	24.2	16.2

[a]Includes small percentage from wood and waste and also, in the past few years, from nuclear power.

Source: U.S. Federal Power Commission, PRESS RELEASE NO. 16323, as cited in U.S. Department of Commerce, Bureau of the Census, STATISTICAL ABSTRACT OF THE UNITED STATES: 1970, p. 507, and 1971, p. 497.

expected to begin production in the 1970s, with even more coming into operation later. The increasing contribution of nuclear power can be seen in a 1971 Federal Power Commission forecast, displayed in Table 3-4.

Hydroelectric Power

Hydroelectric generation uses waterflows powering turbines that turn electric generators. These waterflows can be in the form of run-of-river or storage

Table 3-4
Forecast of Types of Electric Power Production: 1970, 1980, and 1990 (Percent)

Year	Hydro	Fossil-Fuel	Nuclear
1970	16.3	81.8	1.9
1980	14.4	64.6	21.0
1990	12.1	50.2	37.7

Source: Statement of John N. Nassikas, Chairman, Federal Power Commission, U.S. Congress, House, Committee on Interstate and Foreign Commerce, HEARINGS, POWERPLANT SITING AND ENVIRONMENTAL PROTECTION, 92nd Cong., 1st Sess., 1971, p. 439.

projects. In some experiments, the rise and fall of tides are used. In the run-of-river systems, streamflows are used as received at the plant site; this necessitates relatively constant flow conditions. A storage project, on the other hand, includes capacity to regulate flows for seasonal patterns by water retention in reservoirs during high-flow periods and the releasing of it during low-flow periods.

To the end of June, 1970, the hydroelectric generating capacity of the nation was 54.9 million kilowatts.[3] Approximately 7 percent of this capacity represented pumped-storage facilities, in which water is pumped to higher elevation reservoirs during periods of slack electricity demand and then released to power electric generators during times of peak demand. Hydro facilities are particularly useful as peak-load generators because of their ability to start quickly and make rapid changes in output. Other advantages are low operating and maintenance expenses, no fuel costs, and long life expectancies. In addition, hydro units are normally out of service for repairs only a fraction of the time needed for other types of generating facilities.[4]

The investment costs per kilowatt for conventional hydroelectric plants are considerably higher than for steam-electric facilities, although large cost variations occur.[5] These cost fluctuations depend upon the amount and quality of land which must be flooded, the extent of facility relocation, etc. As noted earlier, however, low operating costs help to offset the large initial outlay. Few hydroelectric projects are constructed only for electric power production. Most of the facilities are part of multipurpose developments that provide recreation, irrigation, navigation, and flood control, along with the generation of electricity. The investment costs for pumped-storage hydroelectric facilities are generally lower than for steam-electric plants.

As noted in Tables 3-3 and 3-4, hydro facilities generate a substantial, though declining, proportion of total electric output. This trend is expected to continue through 1990, although pumped-storage developments will be used increasingly to meet peaking requirements. The main reason for the deterioration in the importance of this method of production is the lack of suitable sites for further development near points of electric consumption. Long-distance transmission of electricity may change this someday. Although developed capacity represents only approximately 40 percent of potential capacity,[6] most remaining sites are in the Mountain States and Alaska, where electricity is generally in plentiful supply. In the Middle Atlantic States, where new generating facilities are needed, nearly all potential sites are already in use.

Conventional Steam-Electric Power

Steam-electric facilities burn fossil-fuels in the form of oil, natural gas, or coal to convert water to high pressure steam. This steam then is used to power turbines, which rotate electric generators. As noted in Table 3-3, this is currently the most widely used method for producing electric power, although it will decline somewhat in importance over the next two decades as power companies add additional capacity in the form of nuclear facilities. Although oil and natural gas have become relatively more important as a source of fuel in new facilities in recent years, coal is still the primary fossil-fuel used in producing steam and will keep its dominant position through the foreseeable future. Electric utilities are presently using 300 million tons of coal per year and, even though this is expected to increase to more than one billion tons annually by the year 2000, estimated United States recoverable reserves of over 1,6000 billion tons[7] should prove adequate for many decades. This abundance of coal supplies is in contrast to oil and natural gas in which the United States is no longer self-sufficient on a sustained basis. Although coal availability recently has been a problem area for many electric utilities, much of the blame can be traced directly to the utilities, since they failed to make long-term contracts with coal mining companies in anticipation of large increases in nuclear capacity. In addition, in 1969, the export market for coal improved substantially and labor surpluses at mines turned to shortages, while coal-carrying hopper cars also were in short supply. This was at the same time that nuclear construction delays became evident. These factors combined to cause fuel shortages for electric power companies in many areas of the country, with resulting low reserve capacities and brown-outs. Despite these current problems, coal still remains the foremost fossil-fuel for the future production of electricity.

Nuclear Power

The expansion in electricity output over an intermediate-term of twenty to thirty years projects an increasing use of nuclear-fueled reactors. These power plants are similar to fossil-fueled plants in that heat is used to convert water to high pressure steam, which powers turbines and electric generators. The difference is that, rather than using oil, coal, or natural gas to produce the heat, nuclear energy is employed. Two major types of reactors are presently in use: pressurized-water reactors and boiling-water reactors. Simple diagrams of these power plants are shown in Figure 3-2. The pressurized system is used in approximately 60 percent of the light-water reactors. In this type of system, the primary water coolant is separated from the turbine by a heat exchanger. Water in the primary loop, heated to steam by energy released in the reactor core, transfers its heat to the secondary loop through the heat exchanger. In the secondary loop, steam, which is used to power the turbine, is then converted back into water in the condensing unit and the process is repeated. In the boiling-water process, no heat exchanger is used, and steam from the nuclear reactor is used directly to power the turbine.

The primary disadvantage of both the pressurized-water and the boiling-water processes is the inefficient use of uranium fuel. Each of these reactors effectively utilizes only uranium-235, which comprises less than 1 percent of natural uranium. Should these two types of reactors be the only methods of generating nuclear energy, estimated low-cost uranium resources will be depleted in less than 30 years.[8] The Atomic Energy Commission has developed a supply curve for uranium shown in Figure 3-3. Although fuel costs are a lower proportion of total costs in nuclear installations than in fossil-fuel facilities, a continuation of inefficient uranium use would affect the cost of electric generation with the use of present light-water reactors. It is estimated that, for each dollar increase per pound of uranium, there is a 0.06-mill increase per kilowatt-hour of electricity.[9]

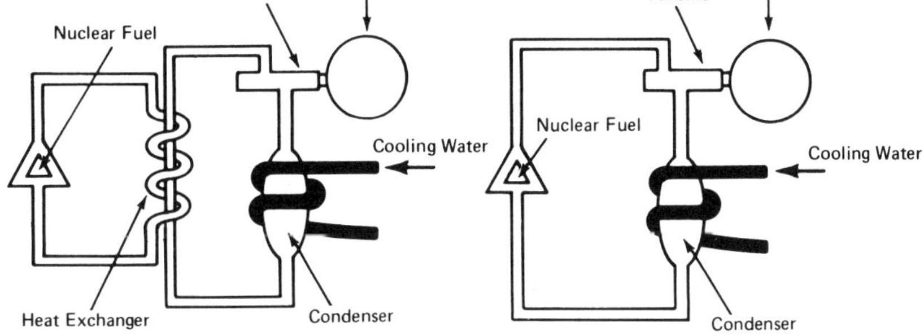

Figure 3-2. Pressurized and Boiling-Water Nuclear Power Plants.

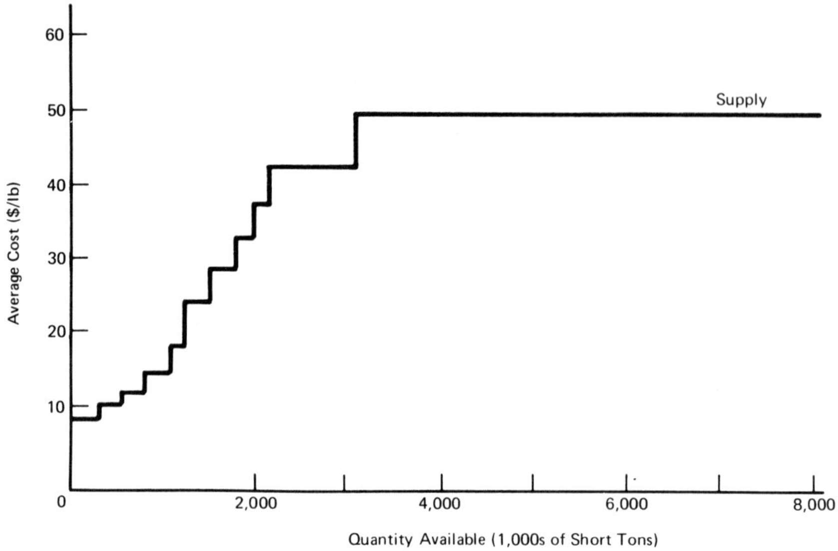

Figure 3-3. Projected Supply of Uranium. Source: P.M. Murphy, R.S. Palmer, and D.B. Sherer, "Incentives for the Development of the Fast Breeder Reactor," *Public Utilities Fortnightly*, LXXXIX, No. 7 (March, 1972) p. 27.

Hence, should low-cost uranium be consumed and the price forced up to $20 per pound from the present $8, the cost of electricity would rise by over 0.7 mill per kilowatt-hour. This is quite a significant increase, considering that the average price to ultimate consumers in 1968 was 1.54 cents per kilowatt-hour.[10] Without more efficient reactors, the projected nuclear plants to be built by the year 2000 in the United States would raise the price of uranium into the region of $30-$50/lb.[11]

The generally agreed-upon answer for a more efficient use of uranium fuel is the liquid-metal fast breeder reactor. Although some development has taken place on gas-cooled fast breeders, most research has centered on the LMFBR. Rather than using scarce U-235, the liquid-metal reactor is fueled with a mixture of plutonium and uranium-238, the latter comprising more than 99 percent of natural uranium. In the reaction process, the breeder actually produces 50 percent more plutonium than it uses. This method will ensure that the nation's uranium resources are stretched to thousands of years. The fast breeder is currently in the experimental stage; its use in significant proportions is not anticipated until the middle or latter 1980s.[12]

Alternative Methods of Electric Power Production

Fusion. A nuclear development much further removed in time than the fast breeder is the use of fusion to produce energy for electric production. Unlike fission–power which generates energy by splitting atoms–fusion power fuses nuclei of some element together. This is the same process that heats the stars (and sun) and powers the hydrogen bomb. The problem has been in man's containing the reaction. Fusion requires extremely high temperature (on the order of 100 million °K) to begin and sustain a reaction. The major difficulty is that the hot fuel gas must be isolated from contact with the reactor walls. This is currently being attempted through the use of magnetic fields. The advantage of fusion power is that light elements, rather than uranium, will be used as fuel. Deuterium, the heavy hydrogen isotope, is the primary fuel and it exists–in the oceans, for example–in quantities sufficient to last for thousands of millions of years.[13] Work in this area began over two decades ago and General Electric and Standard Oil of New Jersey are currently financing further research at the University of Rochester.[14] Although scientists knew about fusion before they knew about fission, there is little possibility that fusion power will be adequately developed for use before the end of this century.

Gas Turbines and Diesels. A common method of generating electricity, when only a limited amount of output is needed, is diesels and gas turbines. These operate directly with generators to produce power and, although sometimes used to meet small base loads, they are generally used for peaking requirements or as backups in case normal sources of supply fail. While fuel costs are usually high, the low fixed cost and labor charges, combined with the ability to start and stop quickly, make gas turbines and diesels satisfactory in limited application.

Magnetohydrodynamics. Proposals have been made for development of a system to obtain electricity from conducting fluids passed through a magnetic field. This process needs no turbines or generators, as gas is converted directly into electrical energy. In addition to high efficiencies, this type of generation promises the capability of use in large-scale power stations.[15] In 1967, Avco Corporation, in conjunction with a number of New England utilities, began research on using MHD for emergency and peaking roles. Avco projected that with significantly increased funding, MHD could become a substantial producer of electricity in the 1980s. Because of its low priority at present, it does not appear that this method of production will become technically feasible on a large-scale basis within the foreseeable future.[16]

Geothermal Power. Geothermal energy is the heat emitted from the earth's magmatic core. This heat turns water to steam and, in electric power application, the steam then is used to power turbines and generators. In many locations, this steam or hot water is naturally available at the surface of the earth. The geysers and hot water pools at Yellowstone National Park are examples of naturally occurring geothermal energy on the earth's surface. In most cases, however, geothermal energy is not so readily available. To take greater advantage of this potential source of energy, the Atomic Energy Commission has proposed in its Plowshare Program to use underground nuclear explosions to provide access to the geothermal heat sources. Under this plan, explosions would be used to fracture underground rocks so that water could be piped into the fracture and converted to steam.

A number of problems accompany geothermal power plants. Owing to the low steam temperature (80 to 200 °C), special turbines must be used.[17] Probably the greatest handicap is that minerals may collect in the steam well shaft, necessitating redrilling. This would involve considerable extra expense. Geothermal power plants are currently in operation in Iceland, Italy, Japan, Mexico, New Zealand, the U.S.S.R., and the United States, but these provide only a very small fraction of world electric output.[18] The largest plant now operating in the United States is the Geysers installation in California, which is of 80-megawatt size. This compares with 1,000 megawatt fossil-fuel and nuclear units currently being built. Total worldwide installed power, using geothermal steam, was 678 megawatts in 1970.[19] With present technology, geothermal energy is economically competitive with other methods of producing electric power,[20] and the simultaneous capture and sale of minerals, taken from the steam, offsets costs further.

Solar Power. A number of different plans that use energy from the sun in the generation of electricity have been proposed. Practical application has taken place in a few instances, with the space program probably being the most successful. The basic problem in the collection of solar energy is that, although the total amount of solar energy striking the earth is very large, the amount on any given portion of the earth is small and intermittent. To help remedy this difficulty, some scientists have proposed that collecting "farms," many square miles in area, be constructed in deserts. A more acceptable approach envisions the orbiting of satellites, which could collect solar energy, convert it to electrical energy, and transmit it to a receiving station on earth by microwave. Because of the inefficiencies and high costs, it is unlikely that solar energy will be used to any great extent in the near future.

Fuel Cells. Although the fuel cell was discovered over a century ago, modern fuel-cell technology did not originate until the 1930s.[21] Basically, fuel cells directly convert the chemical energy of a fuel, such as hydrogen, to low-voltage direct-current electricity. For increased voltages, multiple cells can be used. Experimental cells have been used to power automobiles, tractors, and one demonstration home, but as yet widespread practical application has not taken place. Thirty-two utilities, together with Pratt & Whitney and the Institute of Gas Research, are currently working on a program to develop a natural gas fuel cell, but the funding level still is relatively low.[22] The fuel cell exhibits promise in limited applications, where large amounts of power are not required, but it is not expected to replace generation by central power stations.[23]

Although a number of additional methods of electric power production, such as thermoelectric and thermonic generation, are in the experimentation stage, few show much promise of becoming usable in large and widespread application in the foreseeable future. Geothermal power will continue to be increasingly exploited, although at a limited pace, and fuel cells might become economically competitive when only small amounts of power are required. It appears that over the next thirty to forty years, the large bulk of increased electric needs will be supplied by nuclear-fueled steam plants. If technological improvements are made as currently expected, the liquid-metal fast breeder reactor will begin replacing conventional nuclear plants during the latter part of the 1980s. In addition, fossil-fuel plants (especially coal) will continue to be an important method in the production of electric power.

The Transmission of Electrical Power

As electric generating plants increase in both number and size, new transmission facilities must be constructed. Whereas generating stations once were located near major consuming points, improvements in long-distance transmission of electricity have allowed generation near the fuel sources and away from populated areas. This has been particularly important in the case of nuclear reactors and will be discussed in more detail in a later chapter.

As displayed in Table 3-5, the trend in transmission is toward higher voltage lines to obtain increased transfer capability. Whereas the highest voltage lines in the 1950s were on the order of 350 kilovolts, now 500-kilovolt lines are becoming common, and wide use of 765-kilovolt lines is expected in the future.[24] Research currently is being undertaken on lines in the range of 1,000 to 1,500 kilovolts.[25]

The electric utility industry distinguishes lines used for transmission and lines used for distribution. The former generally include the high-voltage lines of Table 3-5 which transfer electrical energy from generating plants to major load centers, while the latter are used to carry electricity to individual customers. Distribution facilities include lines to customers, low-voltage primary and secondary lines, and transformers to reduce transmission voltages.

Since distribution facilities account for nearly 40 percent of the total

Table 3-5
Net Circuit Miles of High Voltage Lines Placed in Service Each Fiscal Year by Nominal Voltage Classes (Private, Public, and Co-Op Ownership)

Year	69 kv	115 kv	138 kv	161 kv	230 kv	345 kv	500 kv	765 lv	800 kvdc	Total
1970	4,192	1,778	2,099	1,393	3,431	3,069	1,150	191	845	18,148
1969	3,695	2,362	2,031	899	2,506	2,256	1,285	0	0	15,034
1968	1,876	1,919	2,134	391	2,483	1,738	2,569	0	0	13,110
1967	2,440	2,963	2,832	896	4,438	1,391	1,117	0	0	16,077
1966	3,506	1,675	628	802	2,693	2,508	933	0	0	12,745
1965	3,249	2,027	1,913	1,251	2,744	629	283	0	0	12,096
1964	2,652	2,195	1,675	359	1,975	270	0	0	0	9,126
1963	4,280	2,125	1,593	985	2,609	418	0	0	0	12,010
1962	3,235	2,200	1,408	288	1,840	686	0	0	0	9,657
1961	3,999	1,661	1,804	615	1,700	399	13	0	0	10,191

Source: U.S. Federal Power Commission, 1971 ANNUAL REPORT (Washington: U.S. Government Printing Office, 1972), p. 15.

investment costs of electric utilities,[26] technology improvements in this area are especially important. Although improved transmission and distribution systems have allowed greater amounts of power to be carried greater distances at improved efficiencies, higher rights-of-way costs, transmission losses,[27] and a desire by many people to use underground lines, have stimulated interest in a number of possible transmission developments.

The main problem of underground high-voltage transmission lines is the heat buildup which occurs around the lines, causing transmission losses. Underground lines usually are surrounded with soil having high thermal conductivity (clay) to disburse a maximum amount of heat, or the cable can be enclosed in a cooling system to improve its load-carrying efficiency. The current solution to the heating problem is to wrap conductors in oil-impregnated paper and enclose them in oil-filled pipes. Since the resistivity of conductors is reduced at lower temperatures, proposals have been made to refrigerate lines well below ordinary soil temperatures.[28] For example, at 20 °K, the boiling point of liquid hydrogen, the resistivity of aluminum or copper is reduced by a factor of 500.[29] Although the refrigeration units required to cool the lines would be costly, the reduction in transmission-line losses may more than compensate for the additional expense.

A more ambitious plan envisions power stations equipped with electrolyzer plants to convert energy into hydrogen by water electrolysis.[30] The hydrogen would be carried in high-pressure underground pipelines and reconverted to electricity in small power stations, using atmospheric oxygen. Alternatively, the hydrogen could be used as a fuel to supplement today's dwindling supply of natural gas.

While transmission and distribution facilities will continue to be improved, more revolutionary proposals probably will not come into widespread use during the next two to three decades. Hence, conventional-type systems, with their inherent inefficiencies, will continue to provide the bulk of transmission and distribution systems.

The Economics of the Electric Power Industry

The price of delivered electricity is determined by regulating commissions, and is supposed to allow electric power companies to earn a "fair return" on their investment. Although this fair-return concept is interpreted by various commissions somewhat differently, the regulation posed no serious problems so long as aggregate generation, transmission, and distribution costs declined, permitting rate reductions to consumers. Even after costs began increasing somewhat in the 1960s, the average price per kilowatt to ultimate users declined due to the block-rate system used in pricing electricity. (Under the block-rate system average cost per kilowatt declines as more electricity is purchased, even though the rate structure remains unchanged.)

Electricity prices displayed in Figure 3-4 show that the average declined from almost 3 cents per kilowatt-hour during the early 1930s to 1.55 cents per kilowatt-hour in 1968. During this same period, the consumer price index nearly tripled; so in terms of constant dollars, the price of electricity declined by almost 80 percent. While most regulatory commissions reduce rates without hesitation when cost reductions occur for electric power companies, they are reluctant to grant rate increases, even though costs have changed sufficiently to support higher electric prices. Presumably, once commissions become accustomed to applications for rate increases they will be more responsive to the requests. The Federal Power Commission has projected that the cost of generation, transmission, and distribution will rise from 1.55 cents per kilowatt-hour in 1968 to 1.83 cents per kilowatt-hour in 1990, even without adjustments for inflation.[31] Table 3-6 shows these figures along with adjustments for possible rates of inflation throughout the period.

One of the foremost reasons for the bright future of nuclear reactors as a source of energy for generating electricity is the avoidance of higher fuel costs. The electric power industry believes this is one of the major ways of its holding down rising costs for electricity. Although the nuclear plants have higher initial capital costs, the fuel costs are significantly lower. A cost comparison for nuclear, coal, and oil power plants is shown in Table 3-7. Unfortunately, as interest rates climbed during the late 1960s, the difference in capital costs

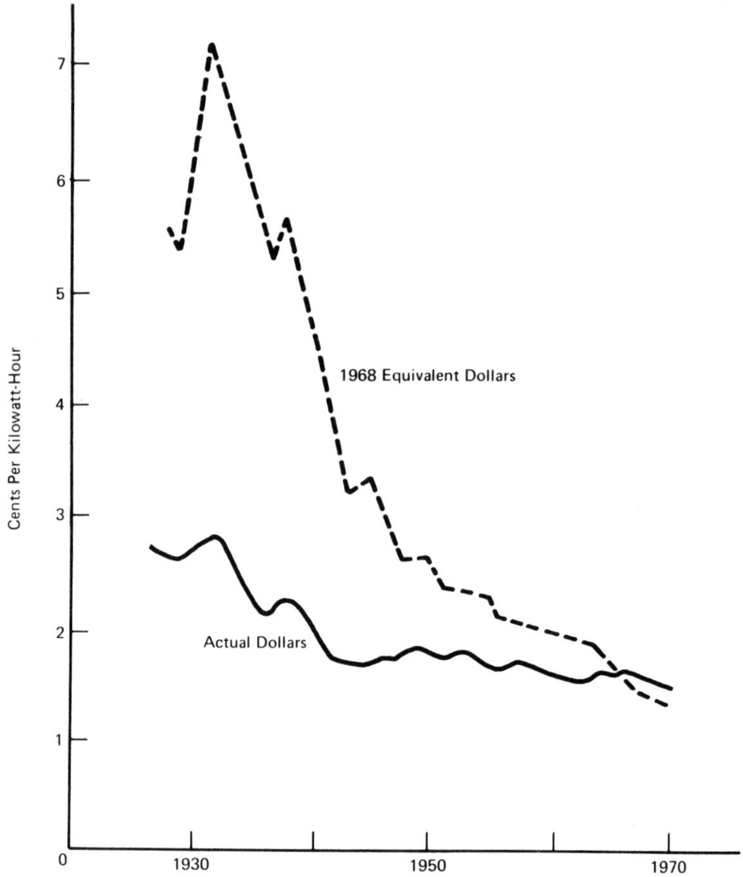

Figure 3-4. Price of Electricity to Ultimate Consumers. Source: U.S. Federal Power Commission, *1970 National Power Survey* (Washington: U.S. Government Printing Office, 1971), p. I-19-2.

became much more important than was at first projected. A doubling of interest rates, from 4 percent to 8 percent, results in extra interest charges of $2 million per year (before taxes) on a $50 million difference in capital costs between nuclear plants and coal-fired plants. The extra interest expense, when comparing nuclear plants to oil-fired plants, is even greater due to the lower capital cost of the latter. This extra interest charge, combined with late delivery dates and environmental problems (which will be discussed in Chapter 4), have made nuclear energy less promising than it once seemed. Also, partially due to the factors mentioned above, expected costs per kilowatt-hour of electricity have risen nearly 300 percent in only eight years.[32]

Table 3-6
Cost of Electricity to Ultimate Consumers: 1968 and Projected 1990

		Cents per kwh	1990 Projected at Various Annual Inflation Rates
U.S. Average–	1968	1.55	
	1990	1.83	1968 Equivalent Dollars
	1990	2.28	1%
	1990	3.51	3%
	1990	5.35	5%

Source: U.S. Federal Power Commission, 1970 NATIONAL POWER SURVEY (Washington: U.S. Government Printing Office, 1971), p. I-19-2.

Table 3-7
Comparison of Large Generating Plant Power Costs (Based on New 1,000 Mw Unit for 1977 Operation)

	Nuclear		Coal		Oil	
	Northeast	Midwest	Northeast	Midwest	Northeast	Midwest
Capital Cost with Wet Cooling Tower but No Sulfur Oxide Control (Millions of Dollars)	335	340	285	295	235	245
Fuel Cost (1970 Avg.) (Cents/Million Btu)	18	18	33	24.5	35.5	50
Power Cost–High Sulfur Fuel (Mills per Kwh):						
Fuel	1.7	1.7	2.9	2.2	3.3	4.6
Fixed Charges	7.9	8.0	6.7	7.0	5.5	5.8
Operations, Maintenance, Etc.	0.3	0.3	0.3	0.3	0.3	0.3
Total	9.9	10.0	9.9	9.5	9.1	10.7

Source: Statement of Edward E. David, Jr., Director, Office of Science and Technology, U.S. Congress, House, Subcommittee on Communications and Power, HEARINGS, POWERPLANT SITING AND ENVIRONMENTAL PROTECTION, 92nd Cong., 1st Sess., 1971, p. 289.

Electric Power Industry Financing

The electric power industry is currently the nation's largest in terms of capital requirements, with the present capital investment of the industry totaling almost $125 billion. However, in the decades of the 1970s and the 1980s, the industry will nearly quadruple this investment in plants and equipment and require total capital outlays on the order of $400 billion to $500 billion.[33] To generate funds to finance these requirements, the industry has relied upon the traditional sources of equity and bond sales, retained earnings, and depreciation. Tables 3-8 and 3-9 show the historical nature of electric utility financing, in both absolute terms and in terms of each source of funds as a percentage of the total sources.

Tables 3-8 and 3-9 show that debt has been the prime source of funds in almost every year. Because of the stability of revenue and earnings growth and the monopoly position of the industry, it is favorably suited to carry large amounts of debt and yet retain prime or near-prime credit-ratings. The greatest advantage of bonds over equity issues, as a method of financial capital investment, is that interest payments are a deductible expense for tax purposes, while dividends are not. Bond issues for the electric power industry nearly quadrupled from $1.3 billion in 1965 to $4.9 billion in 1970; such issues provided more than 40 percent of total funds in all years except 1965.

While earnings stability is suitable to large debt in the industry's financial structure, it has been a hindrance in another respect. Since large earnings increases are abnormal for electric utilities, the price of their common stock is determined primarily by the yield afforded by dividends distributed out of earnings. This yield, adjusted for expected increases in future dividends, must compete with yields on debt instruments. As interest rates climbed to historic levels during the latter 1960s and early 1970s, the prices of the industry's common stocks dropped correspondingly. Whereas price-earnings ratios for the

Table 3-8
Investor-Owned Electric Utilities, Sources of Funds, 1946-1970 (Millions of Dollars)

	External Funds			Internal Funds			
Year	Common Stock	Preferred Stock	Debt	Retained Earnings	Deferred Taxes	Depreciation and Amortization	Total
1946	$ 112	$ (41)	$ 26	$ 67	–	$ 360	$ 524
1950	472	182	632	148	–	483	1,917
1955	415	180	1,080	140	$128	766	1,035
1960	540	164	1,226	379	167	1,181	1,729
1965	379	142	1,261	570	49	1,675	2,294
1966	287	340	2,411	694	43	1,774	5,510
1967	523	465	2,630	591	57	1,894	6,160
1968	623	476	3,161	745	62	2,034	2,841
1969	864	401	3,552	866	97	2,203	3,166
1970	1,795	1,117	4,866	755	111	2,399	3,265

Source: Adapted from U.S. Federal Power Commission, 1970 NATIONAL POWER SURVEY (Washington: U.S. Government Printing Office, 1971), p. I-20-5.

Table 3-9
Investor-Owned Electric Utilities, Sources of Funds, 1946-1970 (Percent)

	External Funds			Internal Funds			
Year	Common Stock	Preferred Stock	Debt	Retained Earnings	Deferred Taxes	Depreciation and Amortization	Total
1946	21.3	–	5.0	12.8	–	68.7	100
1950	24.6	9.5	33.0	7.7	–	25.2	100
1955	15.3	6.6	39.8	5.1	4.7	28.3	100
1960	14.8	4.5	33.5	10.4	4.6	32.3	100
1965	9.3	3.5	30.9	14.0	1.2	41.1	100
1966	5.2	6.2	43.8	12.6	0.8	32.2	100
1967	8.5	7.5	42.7	9.6	0.9	30.7	100
1968	8.8	6.7	44.5	10.5	0.9	28.6	100
1969	10.8	5.0	44.5	10.8	1.2	27.6	100
1970	16.3	10.1	44.1	6.8	1.0	21.7	100

Source: Computed from U.S. Federal Power Commission, 1970 NATIONAL POWER SURVEY (Washington: U.S. Government Printing Office, 1971), p. I-20-5.

common stocks of electric power companies once commanded a premium over the price-earnings ratios of industrial companies, they are now at a large discount. Hence, the market rate of interest is crucially important in the financing scheme of electric power companies; not only do higher rates increase expenses, owing to the large amount of debt carried by most companies, but they also tend to depress the price of common stock, which must compete directly with bonds on a yield basis. The cost of financing electric-utility growth increased significantly in the 1960s and will continue to increase in the future, as utilities replace debt originally issued at 3 percent to 4 percent, with new debt at 7 percent to 8 percent.

Of the internal sources of funds, depreciation is by far the most important. Because of the large amounts of plant and equipment required by the industry, this source has increased consistently in absolute terms, although not as a percentage of total souces. The dollar amount of depreciation cannot change significantly in a short period of time, since, for any given year, new capital investment is only a relatively small percentage of total plant and equipment. Hence, during periods when capital investment is greatly expanding, this source supplies a smaller percentage of total sources, while, during periods of slower growth, depreciation becomes relatively more important. During the latter 1960s, when investment needs expanded rapidly, issues of bonds, common stocks, and preferred stocks also increased significantly to compensate for the deficiency of depreciation allowances.

Since electric utilities pay a greater portion of earnings in dividends than do most other industries (nearly 70 percent compared with 50 percent for all corporate enterprises), and since earnings are restrained by regulating commissions, retained earnings have not provided a very large portion of funds for capital spending.

The foregoing analysis pertains only to investor-owned electric utilities, which undertake nearly 80 percent of the capital expenditures in the industry.[34] Public nonfederal electric utilities rely solely on internal sources and bond issues for financing, since no common or preferred stock can be sold. Bonds have enjoyed wide use by these utilities due to their low interest rates, because of the exemption of interest payments from income for federal tax purposes. Federal electric-utility systems obtain their funds from revenues and the U.S. Treasury. In addition, the Tennessee Valley Authority is authorized to sell bonds and notes in the capital markets.

The trend in future financing depends, to a great extent, upon the future actions of the regulating commissions. Many commissions have been hesitant to allow rate increases, even though higher costs in the industry have supported the need for such increases. Ironically, if commissions delay needed rate increases very long, lower earnings and inadequate coverage of interest payments will lower credit ratings and drive the cost of capital higher. This will add even more pressure for further rate increases.

If capital investment requirements rise, as is forecast by the Federal Power Commission, depreciation will continue to decline in relative importance as a source of funds. Since large increases in bond issues to supply a greater proportion of funds is limited by the large amount of debt already existing in the industry's capital structure, much of the increased financing must come from equity–retained earnings and common stock. The success of these sources, in turn, depends upon regulatory commissions' allowing rate increases.

4 Environmental Effects of Electric Power Production

> Property rights do not extend to the right of individuals and firms to pollute air and water. No one has a license to befoul the amenities and property rights of others. Because no one owns the environment—rather, the public does—there is little to suggest that the public is liable for costs of maintaining its quality. Rather it can be argued that environmental degradation from private activity represents a taking of public property.
>
> Council on Environmental Quality
> August 1971

Most electric utilities have had a difficult time trying to supply power in the quantities demanded by consumers, let alone having sufficient reserves and reliable equipment to prevent temporary shortages.

Aside from the previously examined problems of their financing tremendous amounts of capital equipment and the slow regulating commission reviews of rate-increase requests, electric utilities have suffered major delays in new plant construction because of pressure from environmental groups. While some groups prefer conventional fossil-fuel plants to nuclear reactors, others attempt to delay fossil-fuel plants, thereby forcing electric companies to build more nuclear generating plants. And, as these two factions work at cross purposes, others argue against further hydroelectric development. The result is that power companies many times are forced to follow the path of least political resistance, regardless of the economic or the environmental consequences.

The aforementioned problem has been complicated by a technological advancement—new and better methods of transmission that allow the location of large generating plants far away from the consumers of the electricity. Oregon delays new nuclear plants, hoping the needed electricity will be generated at new hydroelectric facilities in Idaho, while Idaho delays the construction of new hydroelectric plants on the Snake River, hoping construction will begin on nuclear reactors in Oregon. Meanwhile, Southern California passed laws which virtually eliminated new fossil-fuel facilities, knowing that the plants will be built elsewhere. Southern California Edison then participated as a partner in an expansion of the Four Corners plant in New Mexico, transmitting the electricity to its load centers in Southern California by means of 500,000 volt transmission lines.[1] The idea, of course, is to enjoy the amenities that electric power can provide, while letting others suffer from the undesirable effects of its production.

What are the undesirable effects of electric power production? How does the production of electricity affect the environment? It is first necessary to find answers to questions such as these before an intelligent course of action is to be found for producing electric power. Many Americans will continue to demand stringent environmental safeguards and construction delays so long as sufficient electric power is available; but, should a power shortage become apparent and the use of electricity be partially curtailed, electric utilities might well be allowed to expand without regard to the environment. Additionally, because of the capital-intensive nature of the industry, forcing utilities to construct economically-unsound plants to satisfy short-run goals could prove to be a very expensive method of protecting the environment.

Hydroelectric Power

Of the various methods of producing electric power, most people consider hydroelectric generating facilities to have the least overall detrimental effect on the environment. Hydro plants produce no air pollution, water pollution, radioactivity, or solid wastes. With these advantages, one may wonder why more hydro facilities are not built, since only 40 percent of the potential sites currently are being used. One reason is that many of the undeveloped sites are in remote locations where it is not yet economically feasible to produce power. However, at certain sites where it would be desirable—strictly on an economic basis—to build hydro facilities, environmentalists have fought against new dams. Probably the best example of this is the long fight between the Pacific Northwest Power Company and conservationists over the proposed hydroelectric dam for the Snake River in Hells Canyon.[2] Pacific Northwest Power originally filed an application on this proposal in 1954, and the matter is not yet settled. The conservationists realize that the power will be produced in another manner if the company is not allowed to develop the Hells Canyon site, and so they persist in delaying the hydroelectric facility.[3]

The primary environmental effect of a hydro facility generally is the flooding of possibly valuable and scenic land areas. Many of the best sites for power production are in deep canyons that hold aesthetic beauty for many people. These areas—and the wild rivers that flow through them—are disappearing from the American scene, and, once lost, they can never be replaced. John Krutilla argues that conventional benefit-cost analysis on many of these projects understates the present value of the costs. Since the supply of these areas is completely inelastic, the value of the undeveloped site will appreciate over time as more and more people discover the value of its beauty. Also, the present value of benefits derived from such a facility is generally overstated, since new technology will, over time, make the production of electricity by new methods more efficient and the hydro plant more obsolete.[4]

Since most hydroelectric generating facilities are not close to the large areas of electrical consumption, many miles of high-voltage power lines are required for transmission. However, many of the large, new fossil-fuel plants are being built near points where fuel is available, rather than where the power is delivered, so this problem is no longer peculiar to hydro plants.

More recently, environmental alarm has spread over the nitrogen supersaturation in water flowing over the spillways of hydroelectric dams. As excess water, which is not needed for power, spills over a dam, it traps air, which is 79 percent nitrogen, taking it deep into the stilling basin below the dam. The pressure increase forces the gases in the air to dissolve in the water, thereby bringing nitrogen concentrations up to 147 percent of normal.[5] This is a particularly serious problem in the Snake and Columbia Rivers, and the Environmental Protection Agency has estimated that Salmon and Steelhead of the Columbia Basin might be reduced to 10 percent of their present number within three years due to supersaturation of nitrogen.[6]

To offset, at least partially, the harmful effects on the environment of hydro facilities, some environmental improvements also take place after construction of a dam. New recreational activities such as boating, camping, and fishing become available to more people as the area is opened. Of course, to those desiring more wilderness areas, this also might be considered a harmful change. In addition to recreational benefits, many dams provide flood control. This also must be considered a benefit from an environmental standpoint, and could outweigh certain types of environmental harm.

If large reservoirs are formed behind the dams to provide constant waterflow through the turbines, evaporation of the water (and a concentration of minerals) will occur more rapidly. While this might not be a serious problem in rivers possessing very large year-round waterflows, it could prove harmful in river basins that are subject to cyclical flow conditions. Generally, storage projects are preferred to run-of-river dams for rivers with large waterflow fluctuations.

Conventional Steam-Electric Power

Fossil-fuel-fired steam-electric power plants are probably the worst offenders of the environment per kilowatt-hour of power generated. These plants pollute the air, the water, and are aesthetically disagreeable to most people. In addition to solid waste disposal problems, radioactive carbon particulates in fly ash from an older coal plant can produce more radiation exposure than a pressurized water reactor. The problems have been compounded by the fact that, due to economies of scale, utilities have continued to build larger generating units; and, in many cases, they have placed the units close together or in the same plant. Figure 4-1 illustrates the historical growth in plant size, along with projected increases to 1990. Larger plants have become commonplace, even though

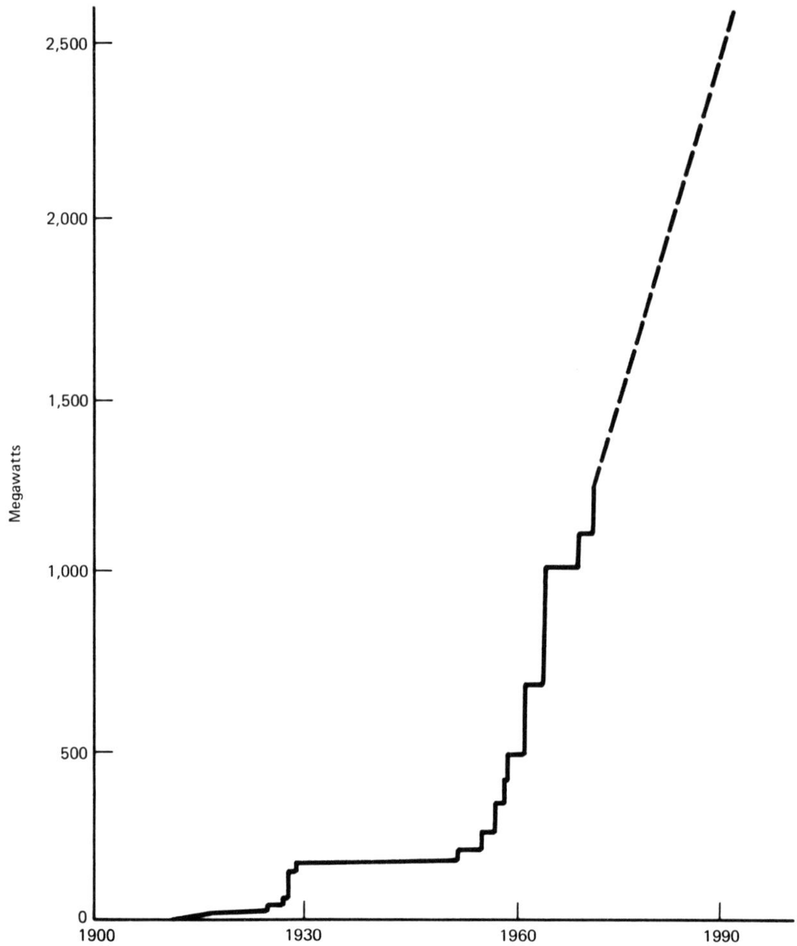

Figure 4-1. Largest Fossil-Fuel Steam-Electric Turbine-Generators in Service: 1900-1990. Source: U.S. Federal Power Commission, *1970 National Power Survey* (Washington: U.S. Government Printing Office, 1971), p. I-5-3.

economies of scale occur mainly on a unit basis, rather than on a plant basis.[7] With the large, widely dispersed power plants, small geographical areas must bear the brunt of the environmental pollution produced by the plants.

Air Pollution

The most obvious residuals emitted from fossil-fuel plants are those which pollute the air. The burning of fossil-fuels results in the emission of hydrocarbons, nitrogen oxides, sulfur oxides, carbon monoxide, and particulate matter. The relative importance of air pollution from fossil-fuel power stations is shown in Table 4-1. The absolute amount—although not necessarily the relative importance—of emissions from electric power plants undoubtedly is substantially greater now than that which is shown for 1966. As can be observed in Table 4-1, fossil-fuel generating plants produced almost 25 percent of particulate matter, 25 percent of the nitrogen oxides, and 50 percent of the sulfur oxides released into the environment in 1966.

The total emissions of plants burning coal, oil, and natural gas are shown in Table 4-2. In terms of sulfur oxides, nitrogen oxides, and particulate matter, coal-burning plants are the greatest polluters, although this is partially due to the fact that coal-fired plants far outnumber both oil-fired and gas-fired facilities.

A relative comparison of the three types of plants is illustrated in Table 4-3.

Table 4-1
Sources of Air Pollution, 1966 (Millions of Tons Annually)

	Carbon Monoxide	Sulfur Oxides	Nitrogen Oxides	Hydro-carbons	Partic-ulates	Total
Motor Vehicles	66	1	6	12	1	86
Industry	2	9	2	4	6	23
Power Plants	1	12	3	1	3	20
Space Heating	2	3	1	1	1	8
Refuse Disposal	1	1	1	1	1	5
Total	72	26	13	19	12	142

Source: Statement of Karl Z. Morgan, Director, Health Physics Division Oak Ridge National Laboratory, U.S. Congress, Senate, Subcommittee on Air and Water Pollution, HEARINGS, UNDERGROUND USES OF NUCLEAR ENERGY, 91st Cong., 2nd Sess., 1970, p. 659.

Table 4-2
Emissions from Coal-Fired, Oil-Fired, and Gas-Fired Steam-Electric Power Plants (Millions of Tons/Year)

	Sulfur Oxides		Nitrogen Oxides		Particulates	
Source	Amount	% of U.S. Total	Amount	% of U.S. Total	Amount	% of U.S. Total
Coal-Fired	13.0	42.7	2.7	15.8	2.26	10.5
Oil-Fired	1.0	3.3	0.3	1.8	0.03	0.1
Natural Gas	(neg.)	–	0.5	2.9	(neg.)	–
Total	14.0	46.0	3.5	20.5	2.29	10.6

Source: National Air Pollution Control Administration, U.S. Congress Joint Committee on Atomic Energy, HEARINGS, ENVIRONMENTAL EFFECTS OF PRODUCING ELECTRIC POWER, 91st Cong., 1st Sess., 1969, p. 811.

Table 4-3
Pollutants from 1,000 Megawatt Coal, Oil, and Gas Power Stations (Millions of Tons/Year)

Pollutant	Coal[a]	Oil[b]	Natural Gas[c]
Particulates	4,950	800	510
Sulfur Dioxide	97,500	58,000	13.5
Nitrogen Oxides	23,000	23,900	13,300
Carbon Monoxide	575	19	–
Hydrocarbons	230	735	–

[a]Based on use of 2.3 million tons of coal, 2.5 percent sulfur, 97.5 percent fly ash removal efficiency.

[b]Based on use of 460 million gallons of oil, 1.6 percent sulfur, no pollution control equipment.

[c]Based on use of 6.8 billion standard cubic feet of gas, no pollution control equipment.

Source: U.S. Atomic Energy Commission, PROCEEDINGS: NUCLEAR POWER AND THE ENVIRONMENT (Washington: U.S. Government Printing Office, 1970), p. 182.

Coal-burning plants are the "dirtiest" in terms of each of the five pollutants, except nitrogen oxides, where oil-fired plants emit slightly more for a similar-sized plant. Although natural gas is the "cleanest" burning fuel, it is not available in sufficient quantities to be used to a great extent in power production. America's reserves of coal far exceed those of oil and gas, and it is coal which must provide most of the fossil fuel for increased electric generation in the future.

Sulfur oxides currently are the worst pollutants emitted from fossil-fuel plants. Not only is their containment difficult (as will be explained in Chapter 5), but they also cause significant damage to health, vegetation, and materials.

Isolation of sulfur oxides to determine their effect on health is a complex problem. A number of studies have been undertaken in this area, and results seem to suggest that most individuals show a response to sulfur dioxide at concentrations of 5 ppm and above. This level varies with individuals, and, for sensitive individuals such as the elderly and those with respiratory problems, 1 ppm can be sufficient to elicit a response.[8] In addition to emphysema, bronchitis, asthma, colds, and pneumonia, sulfur oxides also have been associated with heart disease and lung cancer. Excess deaths occurring in "pollution episodes" in New York in 1953 and 1966, in London in 1952 and 1962, and in Donora, Pennsylvania, in 1948 are generally blamed on high concentrations of SO_2 that existed, even though many other chemicals were present in the polluted air.[9]

In addition to their being a health danger, sulfur oxides also attack physical objects. Sulfuric acid, which is formed as sulfur dioxide is partially oxidized in the atmosphere, deteriorates building stones such as marble and limestone, roofing slate, and mortar. Even without being oxidized, sulfur dioxide damages building coloration, leather, paper, and book bindings.[10]

Nitrogen oxides are formed by the combination of nitrogen and oxygen in the combustion process. There was little concern over nitrogen oxides until the 1950s, when it was discovered that, on exposure to sunlight, they began to form the smog and ozone that has become closely associated with Los Angeles. While smog and ozone are damaging to the eyes and to vegetation, nitrogen dioxide combined with water produces nitric acid, which affects materials, plants, and the lungs.[12] Hence, nitrogen oxides affect health, materials, and vegetation, as do sulfur oxides.

Particulate matter is the most commonly observed pollutant emitted from fossil-fuel power plants. As shown in Table 4-3, particulates are produced by facilities burning coal, oil, and natural gas, although coal-fired plants are by far the worst offenders. (Note that the figure for coal-fired plants in Table 4-3 includes 97.5 percent control.) Particulate emissions depend upon quality of fuel and type of combustion process. Particulate composition from oil-fired plants consists of oil droplets and sulfates, while, from coal-fired plants, it is composed of carbon, silica, alumina, calcium, and iron oxide.[13]

Particulate matter has damaging effects on health, vegetation, and materials, and the results are magnified when it is combined with sulfur oxides. Although the most obvious result of particulate emission is the dirt deposited on objects; it is most damaging to vegetation when it is exposed to moisture, for the dirt solidifies to form hard crusts. Particulates have been related to increased illness and death rates for both high-concentration short-time periods and low-concentration longer time periods.

Particulates reduce visibility and result in what has been called the "icebox effect." This is a decrease in world temperature owing to particulate matter's decreasing the earth's exposure to solar radiation. In addition, greater amounts of rain and snowfall have been observed in areas where particulate emission is heavy, and there is some evidence that these emissions also are responsible for changing the nature of clouds.[14]

Ash proves to be a problem of coal-fired plants, even when it is contained, for it must be disposed of somehow. Prior to 1960, most electric utilities placed the waste in piles near the plants, thereby creating a health hazard when the ash was dispersed by the wind. It is estimated that coal combustion from 1968 to the year 2000 will produce nearly 2 billion tons of waste ash.[15]

Carbon monoxide, while odorless and tasteless, is a highly lethal pollutant. CO cuts off the oxygen supply to the body by replacing oxygen in the red blood cells.[16] Although the amount of this pollutant emitted by electric power plants is small compared with emissions from automobiles, it can nonetheless be dangerous.

The combustion of fossil-fuels has accelerated the generation of carbon dioxide, which normally occurs in nature through decomposition and biological respiration. The net conversion of carbon to CO_2 that normally would occur over millions of years has taken place in less than a century.[17] The increased concentration of about 0.2 percent per year, if continued over a long period of time, could have significant climatic effects on the earth.[18] The buildup of carbon dioxide has been termed the "greenhouse effect," in that CO_2 is transparent to shortwave sunlight, but retards a portion of the longwave heat radiation from escaping the earth, thereby tending to increase the earth's surface temperature.[19] This heating effect, owing to a buildup of carbon dioxide, has been at least partially offset by the cooling effect of the increasing amounts of particulates in the air. In fact, there was a trend toward higher world temperatures from 1900 to the middle 1940s, which has since been reversed to a cooling trend.[20] Possibly the particulate buildup has overcome the increasing amounts of carbon dioxide.

Thermal Pollution

In fossil-fuel steam plants, electricity is generated by having steam expand through a turbine, which in turn drives a generator. Unfortunately, much of the energy released by the burning fuel is converted into waste heat rather than electrical energy. This waste heat is partially dissipated into the air through the stack and through convection and radiation of heat from the boiler. Most of the heat is released into rivers or lakes in the water that flows through the condenser. Currently, the best overall thermal efficiency of fossil-fuel plants is 40 percent, and many of the older plants operate at efficiencies considerably less than this.[21] Hence, only a little over one-third of potential energy from the fuel actually is converted into electrical energy. Further losses occur in the transmission and distribution phases.

A primary requirement of a steam-electric plant is that it be sited near a source of water that is sufficient in size to provide adequate cooling. In a plant that does not attempt to cool the water after use, water is taken directly from the river or lake, circulated through the steam condenser, and returned to its source at a higher temperature. Water used in this once-through cooling process generally averages 15 to 20°F above the ambient water temperature upon its release. The actual increase in temperature depends upon a number of factors, including the amount of water used for cooling purposes. The greater the volume of water passing through the condenser, the less the increase in temperature of that water on a per-gallon basis.

The amount of water withdrawals for condenser cooling is large and is projected to grow rapidly as the number of operating power plants expands. Table 4-4 estimates water withdrawals by type of use for the 1965-2020 period. Withdrawals by steam-electric plants are expected to increase from approximately 30 percent of total water use in 1965 to almost 65 percent in the year

Table 4-4
Estimated Water Withdrawals in the United States (Millions of Gallons Daily)

Type of Use	Used 1965	1980	Projected 2000	2020
Rural Domestic	2,351	2,474	2,852	3,334
Municipal	23,745	33,596	50,724	74,256
Industrial	46,405	75,026	127,365	210,767
Steam-Electric Power:				
Fresh	62,738	133,963	259,208	410,553
Saline	21,800	59,340	211,240	503,540
Agriculture:				
Irrigation	110,852	135,852	149,824	160,978
Livestock	1,726	2,375	3,397	4,660
Total	269,617	442,626	804,610	1,368,088

Source: U.S. Water Resources Council.

2020. The greatest increase will occur in the use of salt water, as more generating stations are located on the coasts. However, these figures can be interpreted in a misleading fashion, since the withdrawal of water does not necessitate its loss to further use. A more accurate indicator is shown in Figure 4-2, in which consumption and withdrawals are differentiated. Consumption indicates that the water is lost to a region's supply once it is utilized. On the other hand, withdrawal includes water consumed and also water which is returned for further use. Hence, the often repeated criticism that "electric power companies will use a certain percentage of the fresh water run-off in the United States by a given year" is not a particularly relevant statement, since use is not the same as one-time consumption.

Unlike air pollution from fossil-fuel steam plants, thermal pollution generally is not considered directly dangerous to public health. The primary problems caused by waste heat are in further use of the water and in the heat's effect on aquatic life.

While the full effects of increased water temperature on aquatic life are not known with great precision, a number of studies have been undertaken in the area.[23] Although water discharged from a plant may be 20°F higher than the temperature of the water source, it is diluted upon reentry—the extent of which depends upon the size of the lake or river used as a heat sink and the volume of heated water released. If the volume of water discharged is extremely small compared with the water used as a receptacle, the damage to aquatic life will be correspondingly small. However, if the volume of heated water is relatively large, serious damage might occur. In addition to the possible adverse effects upon aquatic life, the higher water temperature will increase the rate of evaporation

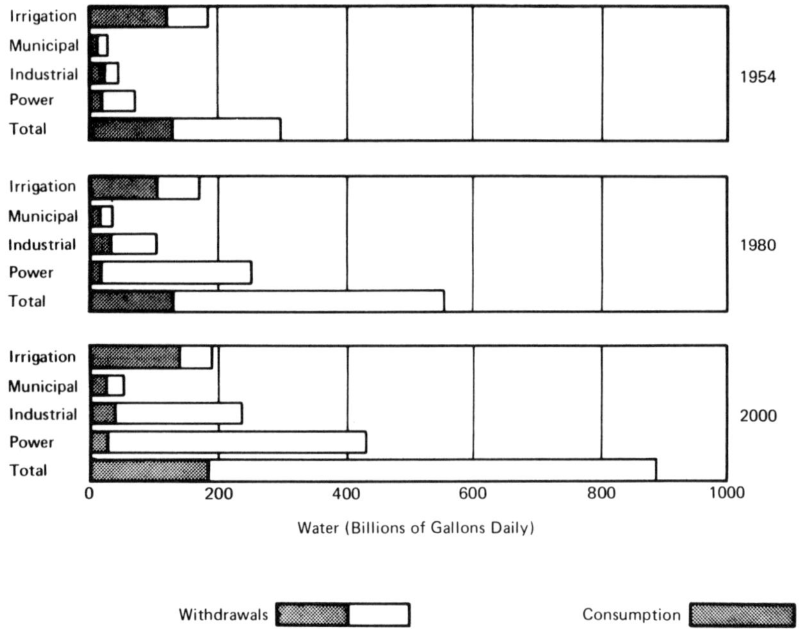

Figure 4-2. Withdrawals and Consumption of Water, 1954-2000 (Billions of Gallons Daily). Source: Statement of Tor Kolflat, Partner, Sargent and Lundy, Engineers, U.S. Congress, Senate, Subcommittee on Air and Water Pollution, *Hearings, Thermal Pollution-1968*, 90th Cong., 2nd Sess., 1968, p. 23.

and reduce the water supply. This can be especially serious where the supply is limited, as in the Southwest; and, if the evaporation is a significant portion of the total supply, a buildup of minerals will occur in the water that remains. While one plant might not raise the water temperature significantly, a number of plants using the same body of water could produce substantial undesirable effects.

Temperature variations affect the growing conditions of algae or phytoplankton. As illustrated in Figure 4-3, temperatures from 68 to 75°F are most favorable for diatoms, 86 to 95°F most favorable for green algae, while above 95°F blue-green algae are dominant. The latter type of algae, which predominates at higher temperatures, is the least desirable food organism for most aquatic life and is the type most responsible for odor and taste problems.[24] In addition, these algae produce water blooms which kill other biota and make water unfit for domestic use.[25] Temperature increases also lead to a reduction in types of organisms. A study of a Pennsylvania stream showed that an increase in water temperature from 80 to 87°F produced 54 percent fewer kinds of invertebrate organisms, while a further increase from 87 to 93°F resulted in an

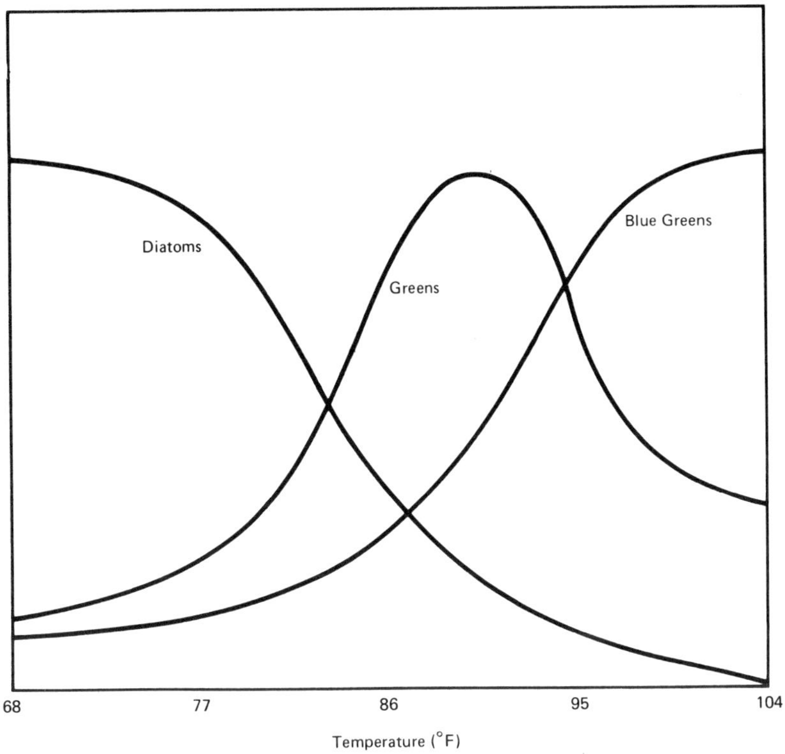

Figure 4-3. Effect of Temperature on Types of Phytoplankton. Source: John Cairns, Jr., "Effects of Increased Temperatures on Aquatic Organisms," *Industrial Wastes* (March-April, 1956), reprinted in U.S. Congress, Senate, Subcommittee on Air and Water Pollution, *Hearings, Thermal Pollution-1968*, 90th Cong., 2nd Sess., 1968, p. 1268.

additional 24 percent loss.[26] The reduction of these organisms, which are a source of food for fish, could have serious repercussions on fish population.

While slight warming of water can provide some beneficial effects—such as faster fish growth and more rapid development of eggs—larger temperature increases lead to a change in the composition of fish population and the possibility of large-scale fish kills. In general, higher organisms are not active in water at temperatures above 95°F and fish of any type are rare above 86°F.[27]

Certain species of fish are more affected by heating of water than others.

Species found in temperate climates usually are able to adjust better to temperature fluctuations than either tropic or arctic varieties, which live in narrow temperature ranges.[28] But, even in temperate climates, water temperature changes which last for long periods of time will alter the composition of fish populations. This occurs not only because the warmer water might prove lethal to some species, but also because various types of fish will no longer be able to reproduce or compete with other types. Generally, it is the less desirable species, such as carp and catfish, which replace more desirable fish like perch and trout as water increases in temperature. While certain species of fish may be able to live and grow in waters of above-normal temperatures, reproduction may be impossible due to the lower temperature requirements for the incubation period. Hence, even though no fish kills occur, over a longer period of time, certain species may disappear because mature fish are unable to breed. Table 4-5 shows the minimum and maximum temperature limits for a survival rate of at least 50 percent for selected freshwater fishes. In addition to fish damage from heated water discharges, some fish also can be harmed when they are sucked from the

Table 4-5
Results of Laboratory Investigation of Minimum and Maximum Temperatures for Certain Freshwater Fishes

Species	Acclimated to °F	Minimum Temperature °F	Minimum Temperature Time (hr)	Maximum Temperature °F	Maximum Temperature Time (hr)
Largemouth Bass	68.0	41.0	24	89.6	72
	86.0	51.8	24	93.2	72
Bluegill	59.0	37.4	24	87.8	60
	86.0	51.8	24	93.2	60
Channel Catfish	59.0	32.0	24	86.0	24
	77.0	42.0	24	93.2	24
Yellow Perch	41.0	–	–	69.8	96
(Winter)	77.0	39.2	24	86.0	96
(Summer)	77.0	48.2	24	89.6	96
Gizzard Shad	77.0	51.8	24	93.2	48
	95.0	68.0	24	98.6	48
Common Shiner	41.0	–	–	80.6	133
	77.0	39.2	24	87.8	133
	86.0	46.4	24	87.8	133
Brook Trout	37.4	–	–	73.4	133
	68.0	–	–	77.0	133

Source: U.S. Federal Power Commission, PROBLEMS IN DISPOSAL OF WASTE HEAT FROM STEAM-ELECTRIC PLANTS (Washington: U.S. Government Printing Office, 1969), p. 19.

rivers or lakes through electric plant cooling systems. Companies generally install screens over the intakes to protect larger fish, but smaller fish still can be pumped through the systems, and some may even become stuck in the screens.

As water becomes warmer, there is an increase in biochemical processes and a reduction in the ability of the water to hold dissolved gases. This can prove damaging to aquatic life since, as the biochemical processes are speeded, oxygen demand also increases. The increased biochemical process also raises the number of bacteria which attack organic wastes to purify water; but, beyond a temperature of approximately 90°F, the bacteria begin to use all of the available oxygen supply and unpleasant odors result.[29] Most fish will die before the odors occur.

While much concern has been expressed about environmental damages due to heated water discharges from electric power stations, beneficial uses also have been discovered. Certain kinds of water organisms that cannot grow in many geographical areas owing to low water temperatures present the possibility of being cultured around steam-electric generating plants. Commercial raising of lobsters and oysters probably represents the most promise in this case. Warm water also is used for irrigation. In addition, steam for desalting and heating currently is being used to some extent, although this use reduces electrical output per unit of fuel input.

While promoters of electric power, including government and power-company officials, like to refer to the waste heat discharges as thermal enrichment, practical and large-scale uses of these discharges are severely limited at the present time. Unless more widespread applications can be employed, thermal pollution is the appropriate term.

Nuclear Power

Nuclear power plants produce no air pollution similar to that emitted by fossil-fuel facilities, but radiation and the disposal of radioactive wastes have presented serious environmental problems. In addition, thermal pollution exists on an even larger scale than for fossil-fuel plants.

Thermal Pollution

Since nuclear facilities produce steam in order to power turbines and generators, they also must use cooling water to condense the steam before recycling it back through the boiler.

While modern fossil-fuel plants produce steam temperatures of over 1,000°F and develop a thermal efficiency of nearly 40 percent, nuclear plants—because of limitations on pressures—have peak steam temperatures of only about 650°F and yield maximum thermal efficiencies of approximately 33 percent.[30] About 10 percent of the waste heat from fossil-fuel plants is discharged through the stacks, unlike nuclear facilities, where nearly all thermal wastes are released through cooling water.[31] The result is that nuclear plants emit nearly 50 percent more waste heat into water than comparably sized fossil-fuel plants. Heat released into the air through stacks generally is considered harmless. The additional heating of water in nuclear steam-electric plants means that the previously cited problems associated with thermal pollution are even more severe than for fossil-fuel plants. Technological improvements in nuclear reactors are expected to increase the thermal efficiencies to levels comparable to modern fossil-fuel units;[32] but, until this actually occurs, the waste heat problem will continue to worsen as nuclear energy supplies ever larger portions of electric output. Figure 4-4 shows a waste heat comparison for current nuclear and fossil-fuel plants and for the nuclear fast breeder reactor, which is expected to become operational in the 1980s.

Radiation

Many environmentalists contend that the principal hazards of nuclear steam-electric plants are the constant release of radioactive residuals, the disposal of radioactive wastes, and the possibility of a loss-of-coolant accident. Most

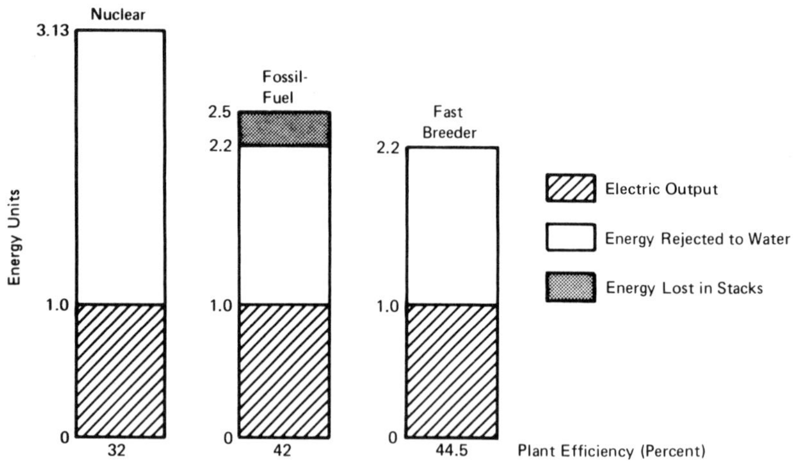

Figure 4-4. Comparative Heat Rejection Potential from Various Generating Plant Systems. Source: Gerald F. Tape, Commissioner, U.S. Atomic Energy Commission, speech presented at the dinner meeting of the Washington Section of the American Nuclear Society, Gaithersburg, Maryland, December 11, 1968.

officials of the Atomic Energy Commission, the Federal Power Commission, and electric utilities using nuclear power (and their suppliers) contend that radioactive release is not a hazard, radioactive waste disposal is no problem, and the chance of a loss-of-coolant accident is so remote as not to be considered relevant.

The main reason that certain environmentalists and scientists have such little trust in statements by proponents of nuclear power is the dual responsibility of the Atomic Energy Commission. While the AEC is charged with the responsibility of protecting the public from the hazards of nuclear energy, it also is charged with advancing the development of civilian nuclear power. Some critics of the Commission contend that the two roles are not compatible and, in order for the AEC to enlarge the scope of its operations, it has been actively promoting the use of nuclear energy at the expense of adequate standards to protect the public. Higher allowable release levels and lower protection requirements result in a more favorable cost comparison for nuclear generating plants relative to fossil-fuel plants.

Another subject that bothers critics of nuclear power plants is the Price-Anderson Act. This Act provides that, for any single nuclear plant disaster, there will be a maximum liability of $560 million in damages on the part of the power company involved. This, in turn, makes it less expensive and easier for the power companies to obtain insurance against such a disaster and is, in reality, a subsidy for nuclear power. What bothers nuclear power critics is not the subsidy, but the unwillingness of insurers to underwrite the total liability for such an accident. A study by the Atomic Energy Commission and the Brookhaven National Laboratory estimated that, for the "worst possible case" of an accident of this type, the total damages would approach $7 billion.[33] Due to the fact that the study was undertaken in 1957, and it considered a plant size of 500,000 kilowatts (today's atomic plants are sometimes twice that size), the damage from an accident of this type today presumably would involve considerably greater damages. If the chance of such an accident is as small as the AEC claims, why is the Price-Anderson Act necessary?

Protection standards for radiation originated over the years in a number of different organizations. Originally, information was supplied to the United States Congress by the National Council on Radiation Protection and Measurements (NCRP) and the International Commission on Radiological Protection (IRCP). However, in 1959, following the controversy on radioactive fallout, the Federal Radiation Council (superceded by the Environmental Protection Agency) was established by President Eisenhower to advise the President on radiation matters.[34] The IRCP established protective standards both for occupational exposure and for whole-population exposure which were accepted by the FRC. Permissible doses for occupational exposure declined from 1.5 rads[35] per week in 1924, to 0.1 rad per week in 1956. For population exposure, the recommended genetic dose is 5 rads per generation (30 years) or 170 millirads per year.[36] The recommendations of the IRCP and the FRC are used by the Atomic Energy Commission as standards to limit radioactive releases from civilian uses of nuclear energy so that exposures do not exceed the guidelines.

Damage to humans from exposure to radioactivity includes both somatic and genetic effects. The former consists of damage such as cancer, cataracts, and impaired fertility, while the latter includes gene mutations and chromosome aberrations, which may cause abortions, stillbirths, and infant mortality.[37] Since these occur naturally, the problem of relating radioactive exposure and the diseases is very difficult.

Originally, the FRC accepted the threshold theory of radiation, which holds that the human body has a tolerance to radiation and can be subjected to certain amounts with no physical harm. In other words, if the radiation emitted by nuclear plants is held below a given level, there will be no ill effects in terms of health damage. A number of scientists still accept this theory; however, a growing number of people in the field hold to the linear theory of radiation damage—that there is a linear relation between the induction of tumors and the size of radiation doses. In the linear theory, somatic effects occur even for small doses, if only on a reduced scale. The scientists accepting the linear theory believe that radiation from nuclear plants should be kept as low as possible. In addition, they contend that the AEC standards are much too lax and considerable health damage would occur if everyone received the maximum permissible dose of 170 millirads per year. One of the foremost critics of the AEC standards is Dr. John Gofman, who contends that if everyone received 170 millirads per year, there would be an additional 30,000 cases of cancer, 150,000 to 1,500,000 extra genetic deaths, and a 5 percent to 50 percent increase in nonfatal, but socially destructive, diseases annually.[38]

Although the AEC recommends that the genetic dose for the total population, on a per capita basis, should not exceed the 170 millirad limit, few nuclear power plants even approach this release level. In most cases, the actual releases are less than 1 percent of the permissible level.[39]

Since the radioactive releases decay and are dispersed as the distance from the nuclear reactor increases, the farther a person lives from the site, the lower the dose he receives. At most plants, persons living near the site boundary receive approximately a 5 millirad dose, while those living within 4 miles receive an average dose of 1 millirad.[40] These doses, of course, consider only releases from the nuclear plants, and not from other sources.

In addition to the releases from a nuclear plant, people are exposed to radioactivity from various other sources. The natural background radiation (including cosmic radiation) ranges from 70 to 200 millirems,[41] depending upon altitude. Outside of the United States, the background radiation ranges up to 500 millirems annually in areas of Brazil, and 1,300 millirems per year in the monazite sand areas of India. Chest X-rays provide a local dose of from 50 to 200 millirems, and living in brick—rather than wooden—homes gives an additional exposure of 50 millirems annually. Somewhat surprisingly, a coast-to-coast plane trip results in a 3 to 5 millirem dose.[42]

Hence, the releases of radioactivity from nuclear power plants are relatively minute, compared with exposures from other sources encountered in everyday living. However, should plants begin releasing radioactive effluents at the maximum limit allowed by the AEC, the comparison will not remain so favorable. For this reason, critics of the standards believe that they should be made more strict, possibly by up to tenfold.[43]

What to do with the radioactive waste that remains after use by the plants continues to be a perplexing problem. Some 30 tons, approximately one-third, of the in-core uranium inventory from each 1,000 megawatt light-water plant is sent off site for reprocessing annually. At the reprocessing plant, further radioactive releases occur, including discharges of krypton and tritium. These plants also are subject to AEC limits and releases typically have averaged 10 percent of the allowable maximum.[44] Of the portion of the spent fuel which cannot be recovered as uranium, plutonium is stored for possible future use, and the remaining radioactive fission products are considered waste.

Until recently, the radioactive wastes that could not be reused were condensed and buried for permanent disposal at a later date. Wastes per ton of uranium used, and which must be disposed of, have declined over the past decade. Whereas a ton of uranium produced 1,500 gallons of radioactive waste ten years ago, it can be condensed to less than 100 gallons today.[45] Recently, solidification of wastes has proved practical. Now all liquid radioactive wastes are concentrated, mixed and heated with other chemicals, and solidified. The solid wastes are combined with glass or ceramic, and then enclosed in metal. Until recently, the Atomic Energy Commission thought abandoned salt mines near Lyons, Kansas, looked like the ideal location for permanent disposal of the solid and liquid wastes that had accumulated from nuclear power plants over the years. The salt mines were believed to be stable, with no flowing water to corrode and reactivate the effluents. Closer investigation revealed that the geological formation at Lyons was not what it first appeared to be, and preliminary drilling indicated water was present in the mines.[46] Currently an alternative to Lyons is being sought, and no solution is yet available for permanent disposal of the residuals.

While there is no apparent danger of a conventional atomic explosion in a nuclear power plant, a break in the cooling system could cause such intense heat from the fuel rods that the reactor core would melt to the bottom of the reactor vessel. This, in turn, might turn the water into high pressure steam and result in an explosion, sending large amounts of radioactivity outside the plant. If no explosion occurred, the fuel mass and supporting structures would continue to melt resulting in a release of virtually all of the gaseous fission products.[47] Nuclear plants include emergency core-cooling systems to provide coolant water to the core should the primary coolant be lost. However, these emergency systems remain virtually untested and their reliability in case of a loss-of-coolant accident is disputed by nuclear power critics.

Should the coolant be lost and the emergency system fail to operate, the results could be of catastrophic proportions. A primary factor that will determine the seriousness of such an accident is the weather at the time of radioactive release. The presence of wind and a temperature inversion would spread the effects over a wider area and affect a greater number of people. With no temperature inversion and an 11.5 mph wind, lethal injuries could be expected over a distance of 2.5 miles, assuming a release of 20 percent of the fission products. With a temperature inversion, 6.5 mph winds, and a 20 percent release, lethal doses would be spread up to 75 miles downwind.[48]

The loss of life and number of injuries would depend upon the distance of the nuclear facility from populated areas. Should such an accident occur in a plant near a large city, the result could be of such proportions that it is difficult to comprehend the damage.

Aesthetic Considerations

Electric power plants probably are no more nor less aesthetically disagreeable than most other industrial complexes of similar size and using similar fuels. Stacks, waste disposal areas, fuel storage facilities, parking lots, and large buildings are common to most plants, whether they produce electricity, automobiles, or guns. However, the physical facilities among the different types of electric generating plants can differ considerably. Coal-fired plants cause the greatest intrusion upon the environment. As shown in Table 4-6, coal plants

Table 4-6
Land Area Requirements for 3,000 Megawatt Coal, Nuclear, Gas, and Oil Power Stations

Fuel	Acres of Land Required	Remarks
Coal	900 to 1,200	Assumes onsite coal and ash storage
Nuclear	200 to 400	
Gas	100 to 200	Assumes pipeline delivery and modest onsite fuel storage
Oil	150 to 350	Assumes adequate onsite fuel storage

Source: Office of Science and Technology, CONSIDERATIONS AFFECTING STEAM POWER PLANT SITE SELECTION (Washington: U.S. Government Printing Office, 1969), p. 11.

consume the largest land area of any of the four types of power plants. Within each area, they require coal shipping and handling facilities, sufficient storage for ashes as they come out of the plant, and large stacks to disperse air pollutants. A 90-day coal supply for a 3,000 megawatt plant would cover 40 acres of land to a height of 40 feet. The requirement for ash disposal would be approximately 300 to 400 acres.[49]

Cooling towers are scenic blights that are just beginning to be seen on a widespread scale. Many plants are installing cooling towers if sufficient water is not available to handle thermal discharges without harming aquatic life. Although these towers are of various designs (as will be discussed in Chapter 5), they all have one thing in common—mammoth size. The size varies with the requirements of each particular site; as plants expand in generating capacity, cooling demands generally increase proportionately. As an example of tower size, an 870 megawatt generating plant near Sacramento, California, scheduled for completion in 1973, will require two natural-draft-type towers, each 425 feet high with a 312-foot-diameter base. While the manager declared the hyperbolic shape of each tower to be "graceful and pleasing to the eye,"[50] it seems that many people probably will view the towers somewhat less favorably.

One major difference in the environmental effects of electric generating plants and other industrial structures results from the different methods of distributing the product. Whereas most manufacturing concerns require trucks, rails, planes, or ships to transfer their products to the consumer, power companies use overhead transmission lines. Industrial manufacturers are not generally subjected to public scorn for their method of distribution, since roads and airports, trucks and planes, are used by others. This is so, even though the environmental effects of this type of distribution probably are more damaging than those of electric transmission lines. However, everyone knows that electric transmission lines are the exclusive domain of power companies.

As electric generating plants expand in size and move away from major load centers, high voltage transmission lines and their huge supporting towers become ever more familiar sights. It is estimated that lines of 69 kilovolts and above currently occupy 4 million acres of land, and by 1990, this will increase to over 7 million acres.[51] Economies of scale exist in transmission lines, as well as in plant size, for electric power companies. The higher the voltage of the lines used, the fewer the lines needed to transmit a given amount of power. While this results in an overall saving in the amount of land devoted to rights-of-way for power transmission, the lines and towers which are constructed are enormous in size. For example, a single experimental transmission tower for a 1,500 kilovolt line is over two-thirds the length of a football field in width and 160 feet in height.[52] If people find the present structures environmentally disturbing, one can only wonder how they will react to those contemplated for future use. As explained earlier, underground transmission lines are not feasible with today's technology.

The pollution aspects of electric power production are considerable and varied. As nuclear plants begin to comprise the bulk of new facilities, there will be an evolution in the importance of different pollutants. In addition to the present emissions of particulate matter, sulfur oxides, and nitrogen oxides, increasing amounts of radioactivity will be discharged. Unfortunately, the environmental hazards of these radioactive emissions, with the exception of large lethal doses, are scantily known. Thermal pollution will be an increasing problem owing to additions of less-thermally-efficient nuclear plants (until introduction of the fast breeder) and the trend toward larger plants. Knowledge of the effects of heated water discharges is lacking. Transmission and distribution facilities will require ever larger sections of land and, although this presents no hazard to health, it places more scars upon the landscape.

5 Control of Pollution in the Electric Power Industry

> Nobody lost any sleep over the fact that competitive circumstances did not permit energy suppliers to price their product for proper land reclamation and the elimination of air and water pollution. Isn't that how we got into this predicament? By serving the public desires, by feeding its appetite for a prosperity purchased with cheap energy? The American people were buying affluence on margin, and their brokers failed to warn them that cheap energy has hidden costs—like polluted air, polluted water, ruptured and abandoned landscapes.
>
> C. Howard Hardesty, Jr.
> Executive Vice President
> Continental Oil Company
> *Newsweek*
> June 12, 1972*

As discussed in Chapter 2, the complete elimination of all pollution from electric power plants is not a socially desirable goal from an economic standpoint. Not only is the technological capacity for the complete control of all emissions currently unavailable, but even in areas where such technology exists, the resources which would have to be devoted to a complete cleanup also could be put to more beneficial use elsewhere. Small releases of certain pollutants seem to cause little or no damage due to the environment's assimilative capacity to absorb the residuals. Even when limited releases do result in damages, such as with radioactivity, the social cost of further emissions control may far outweigh the social benefits derived from the control.

Control of Thermal Pollution

Control devices to eliminate virtually all thermal pollution of water (although not air) are currently available. As in other areas of environmental protection, the expense of control varies directly with the degree of treatment desired. With sufficient controls, and their accompanying costs, water can be returned to its source at the same temperature as it was withdrawn.

Once-Through Cooling

The least expensive method of minimizing the damage from heated water discharges is an alteration of the once-through cooling system. For example, large capacity pumps can be used to circulate increased volumes of water through the condenser, resulting in a smaller temperature rise per gallon of cooling water released. A similar result is obtained when plants dilute heated water with additional cool water pumped directly from the source. If the heated water and source water are mixed in equal quantities before release to the receiving body, the temperature difference will be halved, but the volume of water used will be doubled.

Many of the new nuclear plants are being sited on the coasts so that waste heat can be vented into the oceans. Long discharge pipes can be used to carry the cooling water far out from the shore so that damage will be minimized. However, owing to the corrosive nature of salt water, the cooling equipment must be made of more expensive, noncorrosive materials, and the cost of a discharge pipe for a 1,000 megawatt plant will run up to $5 million per mile.[1]

Another method of altering once-through cooling is to adjust the water release vents to change the flow of heated water into the lake or river. The use of this method varies, depending upon the type of water receptacle available. For example, heated water, discharged so that it affects the entire width of a river, should be avoided since it acts as a "thermal plug," preventing the passage of certain types of fish.[2] If the flow of the warmer water is "guided" down one side of a stream, it is possible for fish to avoid this area and no blockage occurs. However, for a stream which is small, this corrective action might not be possible and more expensive measures would be required.

If the generating plant is situated near a lake, rather than on a river, its intake and discharge points can be located at different water depths. In this way, plant operators can take advantage of the natural temperature variation between water levels (stratification) to reduce the difference in temperature between the discharge water and water at the discharge point. For example, if water is pumped from the same depth as it is vented, and if it is heated by 20°F by the condenser, it will flow into the lake at a 20°F temperature difference. However, if the intake ports are located deeper, where the water is 5°F cooler, there will be only a 15°F difference in temperature between the discharged water and the receiving water at the venting point. Since warm water naturally rises toward the surface, aquatic life below the release depth should suffer minimal damage. The advantages of discharging heated water nearer to the surface are that the affected region is minimized in size and heat dissipation is maximized, since there is a greater difference between surface water and air temperature.[3] If mixing, rather than stratification, is desired, the venting pipes can be located deeper so that the discharged water mixes with cooler bottom water during its rise to the surface.

The most expensive alteration of once-through cooling occurs when discharge canals are employed. The canals are used to cool the heated water partially in transit from the condenser back to the source. The temperature loss, in part, will depend upon the distance the water must flow through the canal. The longer the canal and the greater the cooling effect, the more expensive is this method.

Except in the case of the discharge canal, the heat introduced to a lake or stream is not reduced by the different methods of once-through cooling. Rather, the waste heat is disposed of so as to minimize the possibility of harmful effects in the receiving body of water.

Cooling Ponds

While once-through cooling is the least expensive method of treating thermal pollution, it is also the least effective. A more costly method involves the use of cooling ponds. These shallow ponds lose heat through evaporation, radiation, and conduction (the importance of each depending upon the temperature difference between the released water and the receiving water) after the water passes through the steam condenser. A 1,000 megawatt plant requires on the order of 1,200 to 1,500 acres of pond surface; thus, this treatment is used only when land is relatively cheap.[4] In addition, if the soil structure is such that large amounts of water would be absorbed, the pond must be lined. Should sufficient land be unavailable or too expensive, a spray system can be added to increase the surface area of the water exposed to the air. This extra system can achieve a reduction in the size of the cooling pond by a factor of 20. Since the cooling effect in all cooling ponds occurs mainly through evaporation, water must periodically be added to replace that which is lost to the atmosphere.

One of the major advantages in use of these ponds is the availability of the site for recreational purposes. Relatively small development costs can result in an attractive area for camping, fishing, boating, and other outdoor activities. Also, little operational expense is involved once construction has taken place. The capital expenditures depend almost completely upon the price of land surrounding the plant. If acreage sells in the range of $2,000 to $3,000 per acre, the installed cost of the pond would run from $4 to $6 per kilowatt of electrical capacity for a fossil-fuel plant, and approximately 50 percent more for less efficient (from a cooling standpoint) nuclear plants. If the pond must be lined, the cost could nearly double.

Wet Cooling Towers

When sufficient water flow for once-through cooling is unavailable and land areas for cooling ponds are very expensive, wet cooling towers generally are used by electric power companies. These towers cool heated water through evaporation by spraying it into either forced or natural air currents. The cooled water which remains either is pumped back to the original source or recycled through the condensing unit. In wet towers, nearly 3 percent of the flow is lost in evaporation,[5] so if the process involves continuous recycling between the tower and the condenser, makeup water must be pumped into the system.

Two types of these devices currently are in use—the natural draft tower and the mechanical draft tower, with the latter unit predominating in the United States.

Natural draft towers are built in a hyperbolic shape so that air will naturally flow up through the center section and cool water forms droplets as it falls to the bottom. These are extremely large structures, requiring approximately 4.5 acres of land for a 750 megawatt plant.[6] Although they require higher capital expenditures than forced-air systems, the maintenance costs are less. Chemicals must be added periodically to prevent corrosion, biological growth, and deposits of salt; but, except for these and the pumping cost, few additional operational expenses are involved.

Mechanical or forced-air towers use a fan to blow cooling air through the center section, rather than their relying upon natural air currents. Because of this, the size of the structure required to cool an equal amount of water is reduced, and considerable freedom is allowed in design. The major disadvantage of the forced-air system is that the plant must supply power to run the fans, and this accounts for approximately 3 percent of a plant's electrical output.[7] As with natural draft towers, chemicals must be added to prevent contamination.

Although both types of towers can result in fog and icing from the evaporative process, this is less of a problem with natural draft towers because of the high elevation at which the release occurs. Also, due to the short structure of the forced-air systems, the warm, moist air flowing from the top sometimes is recycled back through the system, thereby reducing the tower's efficiency.

Dry Cooling Towers

The ultimate solution for control of thermal pollution is the installation of dry cooling towers. Owing to the high expense involved, these facilities are used only when water is very scarce and the less expensive wet towers, which mainly cool by evaporation, will not suffice. Since cooling in dry towers does not take place through evaporation, these structures are very large to allow the flow of great quantities of air. However, there is no water loss, and so no makeup water need be introduced into the flow. Figure 5-1 illustrates simple diagrams of wet and dry cooling towers.

Dry towers, like wet towers, can use either forced-air or natural draft for air circulation. Although the mechanical draft type is smaller in size and requires less initial capital expenditure, it necessitates higher operational costs. Dry towers, because they must be exposed to considerably more cooling air than evaporative cooling facilities, require up to 50 percent more land area and can reduce electrical power produced by a plant by up to 20 percent (owing, mainly, to higher condensing temperatures).[8] The investment cost for these devices in a nuclear plant runs around \$27 per kilowatt of installed capacity and, in a

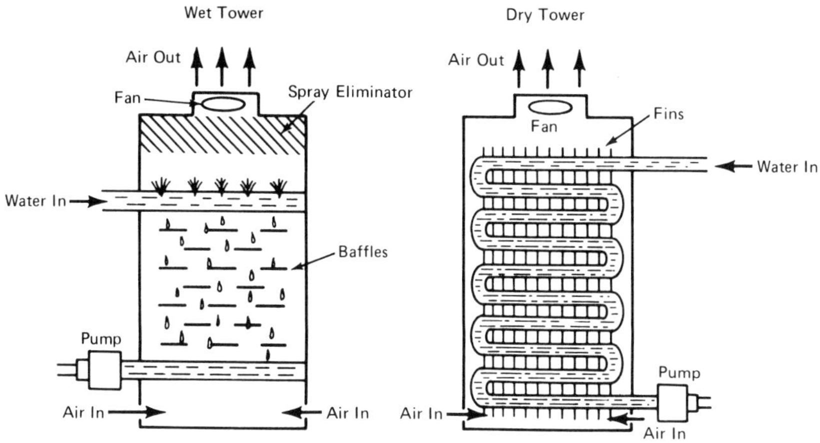

Figure 5-1. Diagrams of Wet and Dry Forced-Air Cooling Towers.

fossil-fuel plant, it approximates \$19 per kilowatt of installed capacity.[9] Relative to evaporative type towers, dry towers can cost 2.5 to 3 times as much per unit of cooling capacity.

Hence, the costs of the various methods of treating thermal pollution vary over a wide range. As illustrated in Table 5-1, the least expensive method of treating waste heat involves an alteration of once-through cooling. This type of

Table 5-1
Comparative Investment Cost of Cooling Water Systems for Steam-Electric Plants

	Investment Cost, \$/kw	
Type of System	Fossil-Fuel Plant	Conventional Nuclear Plant
Once-Through	2-3	3-5
Cooling Ponds	4-6	6-9
Wet Cooling Towers:		
Mechanical Draft	5-8	8-11
Natural Draft	6-9	9-13
Dry Cooling Towers:		
Mechanical Draft	17-21	25-32
Natural Draft[a]	27-30	40-45

[a]Natural draft dry tower estimate from U.S. Federal Power Commission, PROBLEMS IN DISPOSAL OF WASTE HEAT FROM STEAM-ELECTRIC PLANTS (Washington: Federal Power Commission, 1969), p. 15.

treatment will add approximately $3 per kilowatt of installed capacity to fossil-fuel plants and up to $5 per kilowatt to nuclear plants. For a 1,000 megawatt plant, the additional capital expenditure would be $3 million and $5 million for cooling at fossil-fuel and nuclear facilities, respectively. However, if the more expensive dry towers are required, additional investment costs can range upward to $30 million for fossil-fuel-burning plants and $40 million to $45 million for nuclear plants. With the investment cost of a 1,000 megawatt fossil-fuel plant ranging between $150 and $170 per kilowatt and a nuclear plant costing approximately 50 percent more,[10] the water treatment facilities can be a significant addition to the total investment expenditures of the electric power industry.

Table 5-2 shows that fixed plant and equipment charges approximate 50 percent of production costs, and production costs average about one-half of total power costs. Since the investment costs of thermal pollution abatement equipment comprise from 2 percent to 20 percent of fixed plant and equipment costs–depending upon which control equipment is selected–it represents an addition of from 0.5 to 5 percent to the total cost of power. Hence, if wet

Table 5-2
Average Costs of Electricity in the United States, 1968

	Mills/ kwh	Percent of Subtotal	Percent of Total
Power Production Costs			
Fuel	2.47	31.9	
Other Operation and Maintenance	1.34	17.3	
Administration and General	0.23	3.0	
Fixed Charges	3.71	47.9	
Total Production Costs	7.75		50.3
Transmission Costs			
Operation and Maintenance	0.25	12.6	
Administration and General	0.07	3.5	
Fixed Charges	1.66	83.8	
Total Transmission Costs	1.98		12.8
Distribution Costs			
Operation and Maintenance	1.64	28.8	
Administration and General	0.49	8.6	
Fixed Charges	3.56	62.6	
Total Distribution Costs	5.69		36.9
Total Cost of Power	15.42		

Source: Computed from U.S. Federal Power Commission, 1970 NATIONAL POWER SURVEY (Washington: U.S. Government Printing Office, 1971), p. I-19-10.

cooling towers are used to treat water before discharge, an additional $8 million of investment cost is required for a 1,000 megawatt fossil-fuel plant costing approximately $160 million. This $8 million represents 5 percent of total fixed cost of power production and 1.25 percent of the total cost of power. Therefore, the cost of electricity to the consumer can be expected to rise 1.25 percent, from the installation of wet cooling towers. In addition to the installation costs, operation and maintenance expenses will also be necessary. If one assumes that these extra costs are related to tower investment costs in approximately the same proportion as plant operations and maintenance expenses are related to fixed plant costs (1 to 3), they will add an additional 1.7 percent to production costs, or 0.6 percent to total power costs. Hence, evaporative cooling facilities will result in a rise of approximately 1.85 percent in the cost of power.

Should it be necessary to use dry, rather than wet, cooling the cost increase would be significantly higher. Dry tower costs can run over 12 percent of total plant investment expense. Hence, the additional investment cost could result in a 3-percent increase in the total cost of power. In addition, operating costs could add considerably more, especially for forced-air type towers. If the estimate cited earlier by Perry is correct, and mechanical dry towers result in a 20-percent plant power loss, this could add an additional 10 percent to total power costs.

The cost of adequate waste heat control can add as little as a fraction of 1 percent and as much as 10 to 15 percent to the cost of electric power, depending upon which type of abatement facilities are used. In most cases, the less expensive devices will be used in the near future; but, as more generating stations are constructed and the problem becomes more severe, the cost of thermal pollution control can be expected to increase correspondingly.

Control of Air Pollution

Although research efforts are under way on control devices for gaseous residuals from fossil-fuel plants, the only mechanisms currently in widespread use are for control of particulate matter.

Particulate Control

As noted in Chapter 4, coal-fired plants are the primary producers of particulate matter. As the number and size of these plants have increased, public pressure for a reduction in particulate emissions has intensified. The growing strength of protest and the development of control devices are advanced with this pollutant, partially because it is so easily observed, while most other emissions—although damaging—remain invisible. The technological ability to limit particulate emissions was, until recently, far ahead of actual use. This is shown in Figure 5-2. Stricter emissions standards and increased public pressure have resulted in installation of the most modern pollution abatement equipment on many new plants.

While a number of methods exist for the removal of particulates, the two most common devices in use in electric power plants are cyclonic separators and electrostatic precipitators. The former uses centrifugal force to separate larger particulates from flue gases, while the precipitators charge the particulates and attract them to removal screens. Cyclone separators have low efficiencies, since small particulates remain in the gas and are passed through the stack. For this reason, the separators usually must be combined with precipitators if small and

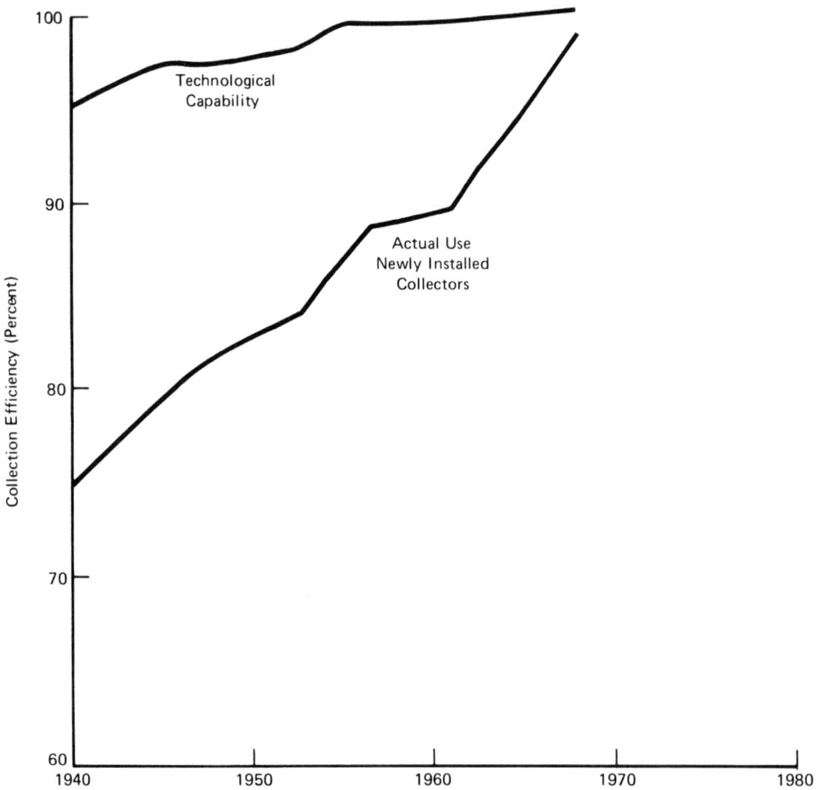

Figure 5-2. Growth in Efficiency Level of Fly Ash Collectors. Source: Statement of Daniel T. Hedden, Chief of Environmental Science, Northeast Utilities Service Co., U.S. Congress, Senate, Joint Committee on Atomic Energy, *Hearings, Environmental Effects of Producing Electric Power*, 91st Cong., 2nd Sess., 1969, p. 1808.

intermediate sized particulates are to be removed. This combination is designed to achieve removal efficiencies of around 99.5 percent,[11] although efficiencies in actual operation have not achieved this level. In theory, the percentage of removal can approach complete control with electrostatic precipitators, in combination with cyclone separators, since an increase in the precipitator size will slow the gases before emission and allow more complete particulate removal. In practice, however, this becomes extremely expensive.

Two additional devices that have been used to a smaller degree are the baghouse and the wet scrubber. The baghouse operates like a vacuum cleaner and catches particulates before they leave the stacks. These are mainly used to catch the finer particulates and so must be accompanied by an additional removal device. High costs are entailed with baghouse use, since the filters must be emptied and periodically replaced. Approximately 20 percent of recovered ash is currently being sold for uses such as concrete production and filling mine voids. These sales help to offset the high maintenance costs of baghouses.

A second device, the wet scrubber, sprays water into the gases. The particulate matter, combined with moisture, is then collected and allowed to settle. This process is efficient at collecting fine particulates; however, it is very expensive and the evaporation cools the gases and thereby reduces dispersion from the stack (the particles fall to the ground closer to the plant).[12]

The major problem with devices that control particulate releases is that, while larger particles are relatively inexpensive to eliminate from stack gases, the smaller particulate matter generally remains. Unfortunately, it is the particulate matter of very small size which causes the most damage. These particulates tend to remain in the atmosphere for long periods of time, thereby contributing to solar radiation reduction. In addition, they are most hazardous to health, since they tend to penetrate to and be retained in the deep lung.[13]

The costs of up-to-date electrostatic precipitators for particulate control range from \$3 to \$6 per kilowatt of installed capacity in new power plants,[14] representing approximately 3 percent of a new fossil-fuel plant that costs \$160 per kilowatt. If maintenance for the precipitator unit runs at about the same rate as maintenance for the whole plant, the installation and maintenance for a modern fossil-fuel plant would increase the total cost of electricity by approximately 1 percent. If it is necessary to add the units to older plants, the costs can be considerably greater. An Edison Electric Institute survey of power companies retrofitting electrostatic precipitators in existing plants found the costs ranged from \$9 to \$24 per kilowatt of installed capacity[15]—nearly four times the cost of installation on new plants. Should retrofitting be necessary, the cost of electric power could increase by 3 to 4 percent.

Should greater removal of fine particles be deemed desirable, baghouse filters can be used in conjunction with precipitators. The baghouse adds approximately 1 percent to the total cost of electricity and makes the expense of approximately 99.9 percent particulate control about 2 percent of total electricity cost for a new plant.

Sulfur Oxides

Unfortunately, control mechanisms for sulfur oxides are not nearly so well developed as devices for particulate control, even though the damage resulting from the oxides probably is greater. As a temporary measure, pending the development of effective sulfur oxide abatement equipment, utilities have had to rely on low-sulfur fuels. These fuels are either naturally low in sulfur content, or else are processed by oil and coal companies before use by power companies.

As the demand for oil of low sulfur content has forced up the price of this fuel in natural form, desulfurization of oil having a higher sulfur content has become more economically feasible. For oil, the reduction of sulfur from 2.5 percent to 0.5 percent costs nearly 70 cents per barrel.[17] Depending upon power plant location, this extra cost results in a 15 to 28 percent higher price for oil at the point of delivery.[18] With fuel charges averaging approximately 16 percent of total electric cost, desulfurization adds from 2.5 to 4.5 percent to the total cost of power.

The problem of removing sulfur from coal prior to combustion is more complex. Coal is partially composed of both organic sulfur and pyritic sulfur. While methods currently are being developed for removal of the latter, elimination of organic sulfur is more difficult, since it is part of the basic coal molecule. Elimination of organic sulfur appears more feasible when coal is chemically processed into a liquid or gas.[19] However, commercial operation of plants to produce synthetic gas from coal is not expected until the early 1980s.[20] If coal of 3 percent sulfur content is composed of one-half pyrite and one-half organic sulfur, the complete elimination of pyrite still would leave the coal with 1.5 percent sulfur content—a percentage which is too high for use in many areas of the country.

The current method of reduction of sulfur oxides is to burn coal which is naturally low in sulfur content; most such coal is found in the western United States. Because of the demand for this type of fuel, it commands a premium price of about \$4 per ton over coal with higher sulfur content.[21] With fuel costs comprising approximately 16 percent of total power costs, and coal selling at a price of \$8 per ton plus freight charges averaging \$3 per ton,[22] the additional \$4 per ton for low-sulfur coal adds approximately 6 percent to the total cost of electricity.

While technology for removing sulfur oxides from stack gases after combustion has been available since the 1930s,[23] commercial use has entailed such high costs that present methods are not now considered economically feasible for abatement.

A number of processes to remove sulfur oxides are currently in the research or demonstration stages. Most of the present efforts are directed toward recovery of a salable residual such as sulfuric acid or elemental sulfur to reduce control costs. However, the cost of reducing these oxides still is expensive. Even with credit for sale of sulfur by-products, costs range from 0.5 to 1.0 mills per kilowatt hour of electricity produced for 90 percent oxide control.[24] With the total cost of electric power at almost 15.5 mills per kilowatt-hour, this would represent a 3.2 percent to 6.5 percent increase in power cost.

Other Pollutants

Little research effort has gone into control of the remaining types of air pollutants from fossil-fuel plants. For example, reduction in emissions of nitrogen oxides has received very low priority, compared with the control of particulate matter and sulfur oxides. It is estimated presently that reductions no greater than 30 to 40 percent will be feasible by 1985.[25]

As power plants have attempted to minimize damage from these pollutants and the remaining portions of particulates and sulfur oxides—at least near the plant site—higher stacks have become very popular. While these do not result in a reduction in emissions, they do provide a greater dispersion of residuals and a decrease in ground level concentrations. The cost of high stacks ranges from $1,000 to $3,000 per foot depending upon a number of variables such as height of the stack, remoteness of the plant, and local material costs.[26] The additional costs of tall stacks would have a negligible effect on the total cost of electricity.

Control of Radioactivity

The cost of reducing radioactive releases is even more difficult to calculate than that of air pollution. For example, even the Atomic Energy Commission seems unsure of how to dispose of liquid and solid radioactive waste. Although most officials believe it eventually must be buried, there is disagreement as to where this should take place. The ultimate answer seems to lie in making the waste harmless through some type of radioactive treatment, but little research has taken place in this area.

Also, it is virtually impossible to estimate the cost of a reduction in the probability of loss-of-coolant accidents and the resulting release of radioactivity. Since companies have experienced difficulty in finding suitable sites for the construction of nuclear plants, more thought has been devoted to the possibility of building plants underground or at sea. Franklyn C. Rogers estimates that the additional cost of locating two 1,000-megawatt units underground would range from $6 to $7 per kilowatt of installed capacity, with this figure taking into account savings from the exclusion of surface facilities.[27] If Rogers' estimate is reasonably accurate, underground nuclear plants would add less than 1 percent to the total cost of electric power.

While radioactive atmospheric and water releases are well below AEC limits, a number of nuclear power critics have suggested that present releases should be reduced. Technology is currently available to allow the additional protection, and equipment costs are not high.

Radioactive releases to water can be reduced by a factor of nearly 100 by a more complete processing of chemical wastes and more complete filtration.[28] The additional operating cost of $50,000 to $70,000 is negligible, in terms of total operating costs, and would add a small fraction of 1 percent to the total cost of power.

Most of the radioactive gas releases occur during condenser air venting. These are partially contained by the plant's holding gases 30 minutes before release; however, a further reduction—by a factor of 20—would take place with the installation of a charcoal absorption system for a longer holding-time, and a device to recombine hydrogen and oxygen before water is held and vented. The capital cost of such a system would range from $750,000 to $1,000,000, adding less than 0.2 percent to the total electric cost.[29]

Aesthetic Improvements

The primary criticism of electric power companies on aesthetic grounds relates to unsightly overhead transmission and distribution lines. This charge has become more relevant as companies have started building large coal-fired plants near mines, because this results in high-voltage transmission lines that run hundreds of miles to distribution and load centers.

Of more than 18,000 miles of transmission lines placed in service in 1970, only a fraction of 1 percent were underground. A total of 2,400 miles of underground lines, 69 kilovolts or more, were in service as of June 30, 1970.[30] The reason for such a small amount of underground high-voltage lines is the extremely high cost of undergrounding. In rural areas, where land for aboveground lines is inexpensive, underground transmission may cost from 10 to 40 times as much as overhead lines; in suburban areas, the ratio may be from 1 to 20 times as great.[31] However, in urban areas, costs may favor undergrounding, and it is in large cities where most of the present underground lines have been placed. Since transmission costs comprise nearly 13 percent of the total cost of power, the undergrounding of high-voltage transmission lines on a large scale appears unlikely, due to the pronounced effect it would have on the price of delivered electricity. The Federal Power Commission's Advisory Committee on Underground Transmission estimated that, if 20 percent of all transmission lines were underground by 1980, consumer electric costs would increase by an average of 20 percent.[32] Costs would increase even faster as more than 20 percent of the lines were put underground, since it becomes a great deal more expensive as voltages increase. Hence, unless vast improvements in undergrounding technology of the type discussed in Chapter 4 occur, there is little likelihood that transmission lines will be placed below ground, except in some suburban areas.

Since only aesthetic considerations are involved, the social cost of overhead transmission lines is impossible to quantify for a justification of underground facilities. Also, since only aesthetic values are included, more importance is being placed upon other types of environmental pollution that cause damage to health.

The cost comparison for underground placement of lower-voltage distribution lines is much more favorable than for high-voltage transmission lines. These have been increasingly used in new housing developments, and, by 1975, 70 percent of new distribution lines are projected to be placed underground.[33]

The effect of underground distribution systems on the total cost of delivered power to consumers depends upon how the extra expenses are financed. Often, when distribution into a housing development is underground, the developer and the power company will share the extra fixed cost in some predetermined proportion. The developer's costs presumably are passed on to the consumers in the form of higher land or housing prices. The greater the proportion of costs the builder pays, the less the direct increase in electricity costs. The Distribution Technical Advisory Committee of the Federal Power Commission has estimated that, with between 70 and 90 percent undergrounding, and all expenses incurred by the electric power industry, the addition to total electric costs will approximate 0.5 mil per kilowatt-hour.[34] At the present cost of 15.5 mills per kilowatt-hour, this results in a 3.2 percent increase in the total cost of electricity. This, of course, does not include undergrounding of existing overhead lines. The Electric Utility Industry Task Force on Environment estimated a cost of $150 billion to convert all existing overhead distribution lines to underground.[35]

Also, it is possible to improve the aesthetic quality of other power facilities. For example, generating equipment and substations can project a more pleasing appearance through proper landscaping and design. Aboveground transmission towers and distribution poles can be structured so as to blend into the surrounding environment or, at least, not be so obtrusive. In addition, changes in line routing can prove more amiable to the environment. Since the expenses can vary so much among facilities, it is virtually impossible to assign any specific cost estimates to the above potential changes.

Total Costs of Pollution Abatement

From the foregoing calculations, some rough cost estimates can be established for controlling pollution in the electric power industry. Exact figures are impossible, since control devices vary so much among geographical areas and from one site to another.

Thermal pollution control depends, to a large extent, upon the availability of

Table 5-3
The Effect of Pollution Abatement on Total Electric Costs (New Plants) (Percent Increase)

Type of Control	Type of Plant: Fossil-Fuel	Nuclear
Thermal Control		
(Wet Towers)	1.8%	2.7%
Air Pollution Control		
Particulates	1.0-2.0	–
Sulfur Oxides	3.2-6.5	–
Radioactivity Control		
Underground Facilities	–	1.0
Aesthetic Improvements		
90% Undergrounding of Distribution Lines	3.2	3.2
Total	9.2%-13.5%	6.9%

Source: Computed from financial data and cost estimates presented in Chapter 5.

water at the plant. Not only is enough water required to provide adequate cooling, but also sufficient water is needed to absorb heat and yet not significantly harm the receiving water.

Air pollution control and radioactive release abatement depends upon what type of fuel is used, local weather conditions, population density, and a number of other factors.

Table 5-3 shows tentative estimates of how much various pollution control methods would add to the total cost of electric power. The major variable is thermal pollution control. Water availability conditions near a power site could result in less expensive once-through cooling or cooling ponds, or it could necessitate the use of very expensive nonevaporative towers. Undergrounding of nuclear plants was used because it partially solves a major drawback of nuclear facilities–the possibility of explosion.

The 3.2 percent increase in power costs due to underground distribution lines represents a minimum figure for aesthetic improvements. Criticisms of other physical structures–such as unsightly plants, substations, and transmission towers–will result in more direct expenses for the industry and its customers; however, it should be recognized that these direct expenses may be offset, to some degree, by indirect savings to society (as in landscaping).

6 Summary, Conclusions, and Recommendations

There ain't no such thing as a free lunch.

Author Unknown

The basic problem in economics—that resources are scarce relative to human wants—applies to an analysis of pollution abatement, as well as to other goods and services. The environment cannot be treated as a priceless asset, to be protected at any cost. Rather, pollution abatement must be weighed against other needs and alternatives.

Economists have developed a framework of analysis for comparing costs and benefits of pollution control. Unfortunately, much of the information which is necessary to implement the framework in pragmatic problem-solving still is lacking. This is evident when an attempt is made to place a dollar value on the damages from pollution. The costs of control equipment are more precise, but until a great deal more knowledge is obtained on damages, the economic analysis is subject to restrictions and value judgments.

The problem of pollution abatement is especially important in the electric power industry. This is true not only because of the industry's rapid growth—a doubling in the use of electricity each decade—but also because relatively inexpensive electricity, along with the primary sources of energy, is the foundation of an industrial society.

It is not possible to curtail the use of electricity and order people to use some other product, as was done with DDT. With some exceptions, such as residential heating and air conditioning, most of the products which are powered by electricity cannot be powered by other means. It also is fruitless to demand new, pollution-free sources of power as a short-run solution. The time required to plan and install a modern nuclear power plant is approximately 10 years; the period necessary to develop a new technology and bring it into practical application is much greater. A liquid-metal fast breeder reactor (which many people find more environmentally disturbing than conventional nuclear plants) has been under development for many years and still is not expected to be operational until the 1980s. It will not result in significant additions to the total electric supply until years later. Numerous other methods of electrical production, such as solar energy, fuel cells, and nuclear fusion, are not expected to approach any type of operational stage until the end of the century.

Since no substitutes for electricity are available in most uses, and since any alternative method of generation is decades away, the only choice is to institute a rational policy of integrating pollution control into the present methods of producing electricity.

Most research on pollution abatement devices for fossil-fuel plants has been concentrated in the control of particulate matter and sulfur oxides. Particulate control technology now is fairly well established and modern devices are rated to reduce emissions to within a fraction of 1 percent; unfortunately, when the mechanisms are put into actual operation, emissions generally exceed this level. Of the particles which escape, most are of the extremely small size that are most damaging to health. However, the installation of more refined (and more expensive) devices such as baghouses should reduce the emissions of even the small particulates.

Control mechanisms to reduce sulfur oxide gases, while currently available, are so expensive that they are not economically feasible under today's price structure. Although a large amount of research has been devoted to the development of less expensive reduction methods, control is being accomplished presently by the utilization of fuels low in sulfur content. As new processes are developed to remove the oxides from the hot gases, they will be used in place of low-sulfur fuels. However, older plants probably will find the initial capital expenditures so high that low-sulfur fuels still will be preferred.

A reduction of heated water discharges is accomplished by various methods, depending mainly upon the nature of the receiving body of water. If water is very scarce, dry towers can be installed so that no cooling water is lost through evaporation. In most cases, however, wet towers, cooling ponds, or some change in once-through cooling will suffice. Because of the large number of new plants which will be built in the future, the lack of new sites with suitable water facilities undoubtedly will necessitate more refined and expensive water treatment facilities. Conventional nuclear plants, which require 50 percent more cooling water than modern fossil-fuel plants, will tend to locate near the oceans, where they can take advantage of less expensive cooling facilities.

While releases of radioactivity into water and the air concern a large number of people, the slight possibility of a loss-of-coolant accident, combined with a malfunction in the emergency cooling system, is a more serious current problem. An occurrence such as this would bring the building of new nuclear plants to a halt and would severely cripple the nation's power-generating capability. Very little research has been conducted on the reliability of emergency cooling systems, and the Atomic Energy Commission has just recently formalized plans for testing the emergency system in a scale model reactor. In addition, some method must be found for disposing of radioactive wastes; it is becoming more apparent that little research was conducted on this problem. All of the environmental problems associated with nuclear power will multiply as atomic plants begin comprising an ever larger proportion of new generating facilities.

The need for underground transmission and distribution facilities, and for more pleasant looking generating facilities, is less pressing. Underground high-voltage transmission lines—except over very short distances—are not anticipated until new technologies are developed which significantly lower the cost of undergrounding. Likewise, only a small effort is under way to underground existing lower-voltage distribution lines. However, additions to the distribution network are projected to be mostly below ground.

The estimates in Chapter 5 of a 9 to 13 percent increase in the cost of power from fossil-fuel plants, and a 7 percent increase from nuclear plants, were made under the assumption of a new plant with average pollution control needs. Depending upon where the plant is sited, the cost of pollution abatement could vary considerably. The greatest variable is the requirement for thermal pollution control equipment. If new plants are located near large bodies of water, the costs for treating heated water discharges could be significantly lowered through use of cooling ponds or long discharge pipes, rather than wet cooling towers, as was assumed. Plants sited along large, fast-flowing rivers, such as the Columbia, also can reduce the costs of waste heat disposal. However, as more generating plants begin locating along the same river or on the same lake, more elaborate cooling facilities will be required.

An additional variable cost factor occurs in the control of sulfur oxide emissions. Effective and economical technology for control of this pollutant now is being developed, and costs are not yet firm. As more experience is gained, the range of control costs computed in Chapter 5 will narrow. The expense will be significant, however, even if actual costs approach the lower end of the estimated range.

The base used in this study is 15.4 mills per kilowatt-hour, which is the average cost of delivered power in the United States. In the cost of electric power, the percentage increase due to pollution abatement varies, depending not only upon the actual cost of control equipment, but also upon the cost of electric power to consumers in the region under consideration. For example, power costs in the southeastern portion of the United States are approximately 67 percent as high as power costs in the Northeast. If the same amount of abatement equipment is required in both locations, and if abatement costs are similar, there will be a significantly greater percentage increase in the cost of delivered electricity in the Southeast as a result of pollution control. The increase will be especially pronounced in the Tennessee Valley Authority area, where rates are artificially low due to federal subsidy.

Another assumption used in this study is that industry cost increases that result from the installation of pollution control equipment will be passed forward entirely to consumers in the form of higher prices for delivered power. As noted in Chapter 3, electric rates are set by regulating commissions to allow power companies a "fair return" on investment. For many years, as the industry's cost of producing and delivering electric power declined, regulating commissions got into the habit of reducing the price of electricity to consumers. This was a trend the commissions found difficult to break and most were reluctant to grant rate increases when costs began increasing. As a result, profits of electric utilities began to be squeezed between higher costs and constant prices for their product. By 1971, however, more increases had been allowed as the commissions apparently resigned themselves to the fact that declining costs for power generation are a thing of the past. It is likely, then, that the costs of pollution abatement eventually will be shifted forward to power consumers, even though the companies might have to absorb portions of the costs while rates are adjusted.

As pollution control devices are installed, and as electric rates to consumers are increased to reflect the higher costs to the power industry, the external costs of electric power production will begin to be internalized. Users of electricity will pay a price which more nearly reflects the full costs of production. However, without complete emissions control, society will continue to bear a portion of the costs in the form of damages from pollution. It is doubtful that payments will be made to individuals to reflect the extent of the damage after controls are instituted. Hopefully, though, social welfare should be increased as a result of the control.

Any recommendations concerning electric power production should be closely related to an overall national energy policy. Unfortunately, no conscious national energy policy currently exists. Decisions have been made only when they were forced through an immediate or impending crisis. In view of Atomic Energy Commission subsidies in support of the nuclear power industry, one would presume that it is in the national interest to produce electricity in nuclear, rather than fossil-fuel, plants. But no such statement of national interest has been presented.

With no government energy policy, there is little wonder that the electric power industry has operated with no apparent coordination in the development of capacity to produce, transmit, and distribute the nation's electrical power. This has resulted in fuel shortages, bad siting of plants, black-outs, plant closings, and operations at less-than-efficient capacity. Only when a long-term plan toward energy use is available can a similar plan for electric power production be formed.

Closely related to the foregoing is a void in the coordination of research and development. Without guidance as to what types of fuels are desirable in the operation of power plants, it is difficult to determine in which areas research and development expenditures should be concentrated. At the present time, the Atomic Energy Commission is focusing its attention on development of the fast breeder reactor. However, many people believe the nation's long-run interests would be served better by reliance on fossil-fuel plants over the short-run and concentration on research efforts in nuclear fusion for longer-run power availability. The Gulf Oil Company subsidiary, Gulf General Atomic, has spent large sums developing a high temperature gas-cooled reactor, which seems to have kindled little interest on the part of either the AEC or the electric power industry. Had the government announced its intention of giving priority to the liquid-metal fast breeder at an earlier date, there is the possibility that funds devoted to the gas-cooled reactor could have been spent in a more productive area.

With the lack of research coordination, there is a duplication of effort in various government agencies. An interagency Committee on Electric Research was established in 1971 to coordinate research in all federal government agencies involved in electric power. However, to date, this effort is less than completely successful. For example, the committee requested $1 million for underground electric research, yet the Atomic Energy Commission—which is a member of the committee—independently budgeted $750,000 to initiate work that is nearly an exact duplication of the CER effort.[1]

In addition, research expenditures have been insufficient to cope with the size and complexity of problems associated with production of electric power and its effect on the environment. It is estimated that total industrial development research and advanced development by both utilities and equipment manufacturers are approximately $150 million per year, representing about 0.2 percent of revenues.[2] Federal government support for electric power research and development added another $315 million in 1970, and probably is somewhat more now; but more than 90 percent of this money represents nuclear research and development.[3]

The small amount of expenditures directed toward research and development results, in large part, from the tremendous drain on available capital which the industry must devote to new plant and equipment expenditures. Most electric power companies have experienced difficulty raising sufficient funds for expansion, and the requirements of pollution control equipment have added even more pressure on capital resources. Some utilities have issued so much debt over the past decade that their prime credit ratings are jeopardized because of the low coverage of interest payments.

Recommendations

The analysis presented in this book provides a background for some specific recommendations concerning the electric power industry and environmental control.

1. Most important, the federal government should adopt a comprehensive energy policy incorporating both short-term and long-term perspectives. In addition to the ideas already discussed, this energy policy should make recommendations on the desirability or undesirability of such services as electrical residential heating. There is also the possibility that natural gas, a comparatively clean-burning fuel, should be prohibited from use by electric

power companies and reserved for applications where pollution control equipment is not economically feasible, such as in households. (This is vigorously opposed by the oil and gas industry.)

2. Rate structures used in electric pricing should be reviewed. Most electric power companies price according to a block rate structure, which may include 5 or 6 rate reductions as more electricity is consumed. The rate structure of the Southwestern Electric Power Company, shown in Table 6-1, is typical of utility pricing policy. Regulating commissions allow this type of pricing because it helps to recover the costs of customer services, such as billing, and because it supposedly helps improve plant utilization. Both of these theories are suspect. A case can be made that customer costs should be recovered in higher billing rates on limited electrical usage. The justification for 5 or 6 blocks is less clear. This type of rate structure promotes the consumption of electricity in uses where its desirability is marginal. Examples are the use of electric water heaters, clothes dryers, and residential heating, where other fuels are competitive. The rates, if properly structured, might improve plant utilization over the short-run, but the main result is to increase the long-run growth of the industry.

3. A 1 to 2-percent excise tax should be levied on electric bills, with the proceeds used to support a balanced research and development program. This program could be administered by an independent governmental or private agency. (Society would benefit from the research and, thus, should pay for part of the cost through use of general tax revenues or tax write-offs.) At any rate,

Table 6-1
Rate Structure of Southwestern Electric Power Company

Usage	Price
Summer Rate (May through October, inclusive)	
First 14 kwh	$1.00
Next 26 kwh	0.047 per kwh
Next 110 kwh	0.03 per kwh
Next 850 kwh	0.02 per kwh
All over 1,000 kwh	0.018 per kwh
Winter Rate (November through April, inclusive)	
First 14 kwh	$1.00
Next 26 kwh	0.047 per kwh
Next 110 kwh	0.03 per kwh
Next 350 kwh	0.02 per kwh
All over 500 kwh	0.015 per kwh
With electric space heating	0.011 per kwh

Source: Moody's Investors Service, MOODY'S PUBLIC UTILITY MANUAL (New York: Moody's Investors Service, Inc., 1971), p. 917.

care should be taken to eliminate a duplication of research and development efforts by different groups. Based on revenues of well over $20 billion annually, this tax would produce a fund of $200 million to $400 million. The 1 to 2-percent tax is an arbitrary figure.

4. The dual responsibility of the Atomic Energy Commission should be ended. The job of its setting standards and protecting the public should be transferred to an independent agency, such as the Environmental Protection Agency, to end the "conflict of interest."

5. Rate relief should be granted readily to compensate the power companies for expenses applicable to the installation and operation of pollution control equipment. Power companies have been lax in instituting environmental control, partially because regulating commissions are sometimes years in granting rate increases to compensate the companies for the extra expense. If the industry knew that additional costs would be quickly translated into higher rates, it would be less hesitant to institute control measures. In addition, the increased rates, coupled with a changed block rate structure, would decrease the quantity of electricity demanded for "marginal" uses (where other fuels would suffice) and should stimulate sales of more efficient electrical-powered devices. These devices generally entail higher construction costs than less efficient mechanisms, and so their purchase by consumers is discouraged by higher retail prices. As the price of electricity rises, however, the higher initial purchase price will be offset by lower operating costs. For example, it is possible to improve substantially the efficiency of air conditioning equipment. Also, fluorescent bulbs can be substituted for incandescent bulbs, and heat pumps can be used in place of resistance heating.

6. Advertising by electric power companies, which seeks to increase the demand for electricity, should be ended or at least disallowed as an operating expense for rate-setting purposes. One of the major problems of the industry today is that demand for the product already is expanding at such a rapid rate that utilities are having a difficult time in providing sufficient capacity for its production. Advertising can only aggravate this problem by creating more demand. In addition, promotional practices such as their granting building contractors cash payments or other financial inducements to install electric heating should be prohibited.[4]

7. Research in long-distance transmission of electricity should be accelerated. One of the principal means of reducing the environmental impact of electric power generation involves the proper siting of plants. Improved transmission facilities would allow siting near energy and water sources and away from population centers, yet permit the energy transfer to load centers without large losses of power. Also, larger power plants could be erected, taking advantage of economies of scale.

Notes

Notes

Chapter 1
Introduction

1. U.S. Congress, Joint Committee on Atomic Energy, HEARINGS, ENVIRONMENTAL EFFECTS OF PRODUCING ELECTRIC POWER, 91st Cong., 2nd Sess., 1969-70.

2. U.S. Congress, Joint Committee on Atomic Energy, SELECTED MATERIALS ON ENVIRONMENTAL EFFECTS OF PRODUCING ELECTRIC POWER (Washington: U.S. Government Printing Office, 1969).

3. U.S. Congress, House, Subcommittee on Communications and Power, HEARINGS, POWERPLANT SITING AND ENVIRONMENTAL PROTECTION, 92nd Cong., 1st Sess., 1971.

4. U.S. Office of Science and Technology, CONSIDERATIONS AFFECTING STEAM POWER PLANT SITE SELECTION (Washington: U.S. Government Printing Office, 1969).

5. U.S. Congress, Senate, Subcommittee on Air and Water Pollution, HEARINGS, THERMAL POLLUTION–1968, 90th Cong., 2nd Sess., 1968.

6. Charles B. Wurtz and Charles E. Renn, "Water Temperatures and Aquatic Life," EDISON ELECTRIC INSTITUTE RESEARCH PROJECT NO. 49, reprinted in Subcommittee on Air and Water Pollution, ibid., pp. 1139-1243.

7. Statement of Tor Kolflat, Partner, Sargent & Lundy Engineers, ibid., pp. 22-89.

8. U.S. Congress, Senate, Subcommittee on Air and Water Pollution, HEARINGS, AIR POLLUTION–1970, 91st Cong., 2nd Sess., 1970.

9. U.S. Congress, Senate, Committee on Interior and Insular Affairs, HEARINGS, PROBLEMS OF ELECTRICAL PRODUCTION IN THE SOUTHWEST, 92nd Cong., 1st Sess., 1971.

10. U.S. Federal Power Commission, 1970 NATIONAL POWER SURVEY (Washington: U.S. Government Printing Office, 1971).

11. U.S. Office of Science and Technology, ELECTRIC POWER AND THE ENVIRONMENT (Washington: U.S. Government Printing Office, 1970).

12. U.S. Council on Environmental Quality, THE PRESIDENT'S 1971 ENVIRONMENTAL PROGRAM (Washington: U.S. Government Printing Office, 1971).

13. U.S. Congress, Senate, Subcommittee on Intergovernmental Relations, HEARINGS, INTERGOVERNMENTAL COORDINATION OF POWER DEVELOPMENT AND ENVIRONMENTAL PROTECTION ACT, 91st Cong., 2nd Sess., 1970.

Chapter 2
The Economics of Pollution Control

1. The Environmental Pollution Panel of the President's Science Advisory Committee defined environmental pollution as "... the unfavorable alteration of our surrounding, wholly or largely as a by-product of man's actions, through direct or indirect effects of changes in energy patterns, radiation levels, chemical and physical constitution and abundances of organisms. These changes may affect man directly, or through his supplies of water and of agricultural and other biological products, his physical objects or possessions, or his opportunities for recreation and appreciation of nature." President's Science Advisory Committee, RESTORING THE QUALITY OF THE ENVIRONMENT (Washington: U.S. Government Printing Office, 1965), p. 1.

2. For example, see the debate by Paul R. Ehrlich, John P. Holdren, and Barry Commoner, "The Closing Circle," ENVIRONMENT, XIV, No. 3 (April, 1972), pp. 23-52.

3. Sanford Rose, "The Economics of Environmental Quality," FORTUNE, LXXXIII, No. 2 (February, 1970), p. 120.

4. An externality exists when the actions of one party affect the utility function of another party who has no control over the former's actions. See James M. Buchanan and W.G. Stubbelbine, "Externality," ECONOMICA, XXIX, No. 4 (November, 1962), p. 372.

5. Collective goods or common property resources have two major characteristics—supply is indivisible, and demand is not specific and individual. Benefits are general in nature, and flow to individuals as members of a group.

6. S.V. Ciriacy-Wantrup, "The Economics of Environmental Policy," LAND ECONOMICS, XLVII, No. 1 (February, 1971), pp. 36-45.

7. Senate Bill S. 2770.

8. Marshall I. Goldman, ECOLOGY AND ECONOMICS: CONTROLLING POLLUTION IN THE 70s (Englewood Cliffs, New Jersey: Prentice-Hall, Inc., 1972), p. 22.

9. U.S. Department of Health, Education, and Welfare, DANGER IN THE AIR: SULFUR OXIDES AND PARTICULATES (Washington: U.S. Government Printing Office, 1970), p. 14.

10. Allen V. Kneese and Blair T. Bower, MANAGING WATER QUALITY: ECONOMICS, TECHNOLOGY, INSTITUTIONS (Baltimore: The Johns Hopkins Press, 1968), p. 162.

11. Statement of Thomas L. Kimball, U.S. Congress, Joint Committee on Atomic Energy, HEARINGS, ENVIRONMENTAL EFFECTS OF PRODUCING ELECTRIC POWER, 91st Cong., 2nd Sess., 1969, p. 1498.

12. Walter G. Belter, Division of Reactor Development and Technology, U.S. Atomic Energy Commission, speech presented at the 136th Annual Meeting of the American Association for the Advancement of Science, Boston, December 28, 1969.

Chapter 3
An Overview of the Electric Power Industry

1. John G. Winger, John D. Emerson, and Gerald D. Gunning, OUTLOOK FOR ENERGY IN THE UNITED STATES (New York: Chase Manhattan Bank, 1968), p. 23.

2. U.S. Joint Economic Committee, Environmental Policy Division, THE ECONOMY, ENERGY, AND THE ENVIRONMENT: A BACKGROUND STUDY (Washington: U.S. Government Printing Office, 1970), p. 8.

3. U.S. Federal Power Commission, 1970 ANNUAL REPORT (Washington: U.S. Government Printing Office, 1971), p. 31.

4. U.S. Federal Power Commission, 1970 NATIONAL POWER SURVEY (Washington: U.S. Government Printing Office, 1971), p. I-7-1.

5. U.S. Federal Power Commission, HYDROELECTRIC POWER EVALUATION (Washington: U.S. Government Printing Office, 1968), p. 27.

6. U.S. Federal Power Commission, 1970 ANNUAL REPORT, op. cit., p. 41.

7. U.S. Joint Economic Committee, THE ECONOMY, ENERGY, AND THE ENVIRONMENT, op. cit., p. 33.

8. Manson Benedict, "Electric Power from Nuclear Fission," SCIENCE AND PUBLIC AFFAIRS, ~~XXVII~~, No. 7 (September, 1971), p. 12.

9. Ibid.

10. U.S. Federal Power Commission, 1970 NATIONAL POWER SURVEY, op. cit., p. I-1-34.

11. U.S. Atomic Energy Commission, WHY FUSION? (Washington: U.S. Government Printing Office, 1970), p. 285.

12. For a more detailed analysis see U.S. Atomic Energy Commission, LMFBR DEMONSTRATION PLANT PROGRAM: PROCEEDINGS OF SENIOR UTILITY STEERING COMMITTEE AND SENIOR UTILITY TECHNICAL ADVISORY PANEL (Washington: U.S. Government Printing Office, 1972).

13. R.F. Post, "Fusion Power: The Uncertain Certainty," SCIENCE AND PUBLIC AFFAIRS, XXVII, No. 8 (October, 1971), p. 42.

14. WALL STREET JOURNAL, May 19, 1972, p. 15.

15. U.S. Office of Science and Technology, Executive Office of the President, MHD FOR CENTRAL STATION POWER GENERATION: A PLAN FOR ACTION; PREPARED BY PANEL ON MAGNETOHYDRODYNAMICS (Washington: U.S. Government Printing Office, 1969), p. 5.

16. Statement of Arthur Kantrowitz, Director, Avco Everett Research Laboratory, U.S. Congress, Senate, Subcommittee on Intergovernmental Relations, HEARINGS, INTERGOVERNMENTAL COORDINATION OF POWER DEVELOPMENT AND ENVIRONMENTAL PROTECTION ACT, 91st Cong., 2nd Sess., 1970, pp. 793-820.

17. David Fenner and Joseph Klarmann, "Power from the Earth," ENVIRONMENT, XIII, No. 10 (December, 1971), p. 26.

18. Ibid., p. 21.

19. Joseph C. Horvath and Robert L. Chaffin, "Geothermal Energy: Its Future and Economics," ATLANTA ECONOMIC REVIEW, XXI, No. 12 (December, 1971), p. 30.

20. Alvin Kaufman, "An Economic Appraisal of Geothermal," PUBLIC UTILITIES FORTNIGHTLY, LXXXVIII, No. 12 (December, 1971), p. 30.

21. Terri Aaronson, "The Black Box," ENVIRONMENT, XIII, No. 10 (December, 1971), pp. 10-11.

22. U.S. Federal Power Commission, 1970 NATIONAL POWER SURVEY, op. cit., p. I-9-9.

23. Ibid., p. I-9-10.

24. U.S. Federal Power Commission, 1971 ANNUAL REPORT, (Washington: U.S. Government Printing Office, 1971), p. 15.

25. Ibid.

26. U.S. Federal Power Commission, 1970 NATIONAL POWER SURVEY, op. cit., p. I-14-1.

27. No detailed breakdown of energy loss in transmission lines is available. The longest high-voltage lines currently in operation are a 400 kilovolt dc line and a 500 kilovolt ac line; each runs 850 miles from near Portland, Oregon, to major load centers in California.

28. Donald P. Snowden, "Superconductors for Power Transmission," SCIENTIFIC AMERICAN, CCXXVI, No. 4 (April, 1972), pp. 84-91.

29. Ibid., p. 87.

30. Derek P. Gregory, "A New Concept in Energy Transmission," PUBLIC UTILITIES FORTNIGHTLY, LXXXIX, No. 3 (February 3, 1972), pp. 21-29.

31. U.S. Federal Power Commission, 1970 NATIONAL POWER SURVEY, op. cit., p. I-19-2.

32. Peter J. McTague, G.J. Davidson, R.M. Bredin, and A.A. Herman, "The Evolution of Nuclear Plant Costs," NUCLEAR NEWS, XV, No. 2 (February, 1972), p. 31.

33. U.S. Federal Power Commission, 1970 NATIONAL POWER SURVEY, op. cit., p. I-19-2.

34. U.S. Joint Economic Committee, Environmental Policy Division, THE ECONOMY, ENERGY, AND THE ENVIRONMENT, op. cit., p. 78.

Chapter 4
Environmental Effects of Electric Power Production

1. U.S. Congress, Senate, Committee on Interior and Insular Affairs, HEARINGS, PROBLEMS OF ELECTRICAL POWER PRODUCTION IN THE SOUTHWEST, 92nd Cong., 1st Sess., 1971, pp. 25-26.

2. U.S. Congress, Senate, Subcommittee on Parks and Recreation, HEAR-

INGS, HELLS CANYON-SNAKE NATIONAL RIVER, 92nd Cong., 1st Sess., 1971.

3. U.S. CONGRESSIONAL RECORD, 92nd Cong., 1st Sess., 1971, CXVII, No. 92, pp. 9212-9214.

4. John Krutilla, "Conservation Reconsidered," AMERICAN ECONOMIC REVIEW, LVII, No. 4 (September, 1967), pp. 777-786.

5. Subcommittee on Parks and Recreation, op. cit., p. 356.

6. U.S. Environmental Protection Agency, SUMMARY REPORT: NITROGEN SUPERSATURATION IN THE COLUMBIA AND SNAKE RIVERS (Seattle: Environmental Protection Agency, 1971), p. 1.

7. Charles E. Olson, COST CONSIDERATIONS FOR EFFICIENT ELECTRIC SUPPLY (East Lansing: Michigan State University Press, 1970), p. 41.

8. U.S. Department of Health, Education, and Welfare, Public Health Service, AIR QUALITY CRITERIA FOR SULFUR OXIDES (Washington: National Air Pollution Control Administration, 1969), p. 100.

9. Thomas K. Sherwood, "Must We Breathe Sulfur Oxides?" TECHNOLOGY REVIEW (January, 1970), reprinted in U.S. Congress, Senate, Subcommittee on Air and Water Pollution, HEARINGS, AIR POLLUTION—1970, 91st Cong., 2nd Sess., 1970, p. 183.

10. U.S. Department of Health, Education, and Welfare, DANGER IN THE AIR: SULFUR OXIDES AND PARTICULATES (Washington: U.S. Government Printing Office, 1970), p. 8.

11. Ibid., p. 9.

12. U.S. Federal Power Commission, 1970 NATIONAL POWER SURVEY (Washington: U.S. Government Printing Office, 1971), p. I-11-7.

13. Ibid., p. I-11-8.

14. Vincent J. Schaefer, "The Inadvertent Modification of the Atmosphere by Air Pollution," BULLETIN OF THE AMERICAN METEOROLOGICAL SOCIETY, L, No. 4 (April, 1969), reprinted in Subcommittee on Air and Water Pollution, HEARINGS, AIR POLLUTION—1970, op. cit., pp. 102-109.

15. U.S. Joint Economic Committee, THE ECONOMY, ENERGY, AND THE ENVIRONMENT (Washington: U.S. Government Printing Office, 1970), p. 109.

16. THE AIR FORCE DRIVER, IV, No. 6 (November, 1970), p. 11.

17. E.W. Barrett, R.F. Pueschel, H.K. Weickmann, and P.M. Kuhn, INADVERTENT MODIFICATION OF WEATHER AND CLIMATE BY ATMOSPHERIC POLLUTANTS (Washington: U.S. Government Printing Office, 1970), p. 92.

18. Ibid.

19. Lamont C. Cole, "Thermal Pollution," BIOSCIENCE, XIX, No. 11 (November, 1969), p. 991.

20. Statement of Dr. John H. Ludwig, Assistant Commissioner for Science and Technology, National Air Pollution Control Administration, U.S. Congress,

Joint Committee on Atomic Energy, HEARINGS, ENVIRONMENTAL EFFECTS OF PRODUCING ELECTRIC POWER, 91st Cong., 2nd Sess., 1969, p. 448.

21. U.S. Federal Power Commission, PROBLEMS IN DISPOSAL OF WASTE HEAT FROM STEAM ELECTRIC PLANTS (Washington: Federal Power Commission, 1969), p. 1.

22. Statement of Dr. Clarence E. Larson, Commissioner, U.S. Atomic Energy Commission, Joint Committee on Atomic Energy, op. cit., p. 234.

23. For an excellent bibliography on this subject see, V.S. Kennedy and J.A. Mihursky, "Bibliography on the Effects of Temperature in the Aquatic Environment," Subcommittee on Air and Water Pollution, HEARINGS, THERMAL POLLUTION–1968, op. cit., pp. 471-568.

24. Statement of Donald Mount, Director, National Water Quality Laboratory, Joint Committee on Atomic Energy, op. cit., p. 358.

25. Cole, loc. cit.

26. Mount, op. cit., p. 359.

27. Cole, loc. cit.

28. Joint Committee on Atomic Energy, op. cit., p. 2453.

29. Federal Power Commission, PROBLEMS IN DISPOSAL OF WASTE HEAT FROM STEAM-ELECTRIC PLANTS, op. cit., p. 22.

30. Statement of Arthur Kantrowitz, Director, Avco Everett Research Laboratory, U.S. Congress, Senate, Subcommittee on Intergovernmental Relations, HEARINGS, INTERGOVERNMENTAL COORDINATION OF POWER DEVELOPMENT AND ENVIRONMENTAL PROTECTION ACT, 91st Cong., 2nd Sess., 1970, p. 793.

31. Larson, op. cit., p. 267.

32. Ibid.

33. U.S. Atomic Energy Commission, THEORETICAL POSSIBILITIES AND CONSEQUENCES OF MAJOR ACCIDENTS IN LARGE NUCLEAR POWER PLANTS (Washington: U.S. Government Printing Office, 1957), p. viii.

34. James T. Ramey, Commissioner, U.S. Atomic Energy Commission, speech presented at the Conference on Universities, National Laboratories, and Man's Environment, Chicago, July 28, 1969.

35. A rad is a unit of absorbed dose of radioactivity.

36. Patricia J. Lindor and J. Rotblat, "Radiation Pollution of the Environment," SCIENCE AND PUBLIC AFFAIRS, XXVII, No. 7 (September, 1971), pp. 19-20.

37. Ibid., p. 18.

38. Statement of Dr. John Gofman, Livermore Radiation Laboratory, Pennsylvania State Senate, Select Committee on Nuclear Electricity Generation, August 20, 1970, reprinted in U.S. Congress, Senate, Subcommittee on Air and Water Pollution, HEARINGS, UNDERGROUND USES OF NUCLEAR ENERGY, 91st Cong., 2nd Sess., 1970, p. 1369.

39. Statement of Theos J. Thompson, Commissioner, U.S. Atomic Energy Commission, Joint Committee on Atomic Energy, op. cit., pp. 192-193.

40. Statement of Clarence E. Larson, Commissioner, U.S. Atomic Energy Commission, Subcommittee on Air and Water Pollution, HEARINGS, UNDERGROUND USES OF NUCLEAR ENERGY, op. cit., p. 672.

41. Since radiation from various sources is rated in different units, the rem is used as a standard unit of equivalence. A one rem-dose is considered to be the equivalent of a dose of one rad due to X-, gamma, or beta radiation. See U.S. Atomic Energy Commission, STANDARDS FOR PROTECTION AGAINST RADIATION, reprinted in U.S. Congress, Joint Committee on Atomic Energy, SELECTED MATERIALS ON ENVIRONMENTAL EFFECTS OF PRODUCING ELECTRIC POWER (Washington: U.S. Government Printing Office, 1969), pp. 93-100.

42. Statement of Clarence E. Larson, Commissioner, U.S. Atomic Energy Commission, Subcommittee on Air and Water Pollution, HEARINGS, UNDERGROUND USES OF NUCLEAR ENERGY, op. cit., pp. 672-673.

43. Statement by Linus Pauling, Professor of Chemistry and Nobel Prize Winner, Stanford University, Subcommittee on Air and Water Pollution, HEARINGS, UNDERGROUND USES OF NUCLEAR ENERGY, op. cit., p. 1365.

44. Statement of T.C. Runion, President, Nuclear Fuel Services, Inc., Joint Committee on Atomic Energy, op. cit., p. 1701.

45. Joint Economic Committee, op. cit., p. 113.

46. John Lear, "Radioactive Ashes in the Kansas Salt Cellar," SATURDAY REVIEW, LV, No. 8 (February 19, 1972), p. 42.

47. Ian A. Forbes, Daniel F. Ford, Henry W. Kendall, and James J. MacKenzie, "Nuclear Reactor Safety: An Evaluation of New Evidence," NUCLEAR NEWS, XIV, No. 9 (September, 1971), p. 33.

48. Ibid., p. 34.

49. Office of Science and Technology, CONSIDERATIONS AFFECTING STEAM POWER PLANT SITE SELECTION (Washington: U.S. Government Printing Office, 1969), p. 14.

50. Statement of Paul E. Shadd, General Manager, Sacramento Municipal Utility District, Joint Committee on Atomic Energy, op. cit., p. 1842.

51. Statement of Carl E. Bagge, Vice Chairman, Federal Power Commission, ibid., p. 459.

52. Derek P. Gregory, "A New Concept in Energy Transmission," PUBLIC UTILITIES FORTNIGHTLY, LXXXIX, No. 3 (February 3, 1970), p. 23.

Chapter 5
Control of Pollution in the Electric Power Industry

1. Statement of S. Baron, Vice President, Burns and Roe, U.S. Congress, Senate, Joint Committee on Atomic Energy, HEARINGS, ENVIRONMENTAL

EFFECTS OF PRODUCING ELECTRIC POWER, 91st Cong., 2nd Sess., 1969, p. 1566.

2. Gerald E. Arnold, General Superintendent, Philadelphia Water Department, speech presented at the Annual Conference, Philadelphia, June 18, 1962, reprinted in U.S. Congress, Senate, Subcommittee on Air and Water Pollution, HEARINGS, THERMAL POLLUTION–1968, 90th Cong., 2nd Sess., 1968, p. 84.

3. Statement of Donald R.F. Harleman, Professor of Civil Engineering, Massachusetts Institute of Technology, Subcommittee on Air and Water Pollution, op. cit., p. 179.

4. U.S. Atomic Energy Commission, PROCEEDINGS, NUCLEAR POWER AND THE ENVIRONMENT (Washington: U.S. Government Printing Office, 1969), p. 100.

5. Statement of Tor Kolflat, Partner, Sargent & Lundy, Engineers, Subcommittee on Air and Water Pollution, op. cit., p. 51.

6. Statement of Harry Perry, Research Advisor to Assistant Secretary for Mineral Resources, Joint Committee on Atomic Energy, op. cit., p. 336.

7. SFI BULLETIN NO. 191 (January-February, 1968), reprinted in Subcommittee on Air and Water Pollution, op. cit., p. 11.

8. Perry, loc. cit.

9. Korflat, op. cit., p. 54.

10. U.S. Federal Power Commission, 1970 NATIONAL POWER SURVEY (Washington: U.S. Government Printing Office, 1971), p. I-19-4.

11. U.S. Office of Science and Technology, CONSIDERATIONS AFFECTING STEAM POWER PLANT SITE SELECTION (Washington: U.S. Government Printing Office, 1968), p. 106.

12. Statement of J.L. Shapiro, Environmental Engineer, Bechtel Corporation, U.S. Congress, Senate, Committee on Interior and Insular Affairs, HEARINGS, PROBLEMS OF ELECTRICAL POWER PRODUCTION IN THE SOUTHWEST, 92nd Cong., 1st Sess., 1971, p. 576.

13. Statement of Paul W. Spaite and Robert P. Hangebrauch, National Air Pollution Control Administration, U.S. Congress, Senate, Subcommittee on Air And Water Pollution, HEARINGS, AIR POLLUTION–1970, 91st Cong., 2nd Sess., 1970, p. 176.

14. Statement of Edison Electric Institute, Joint Committee on Atomic Energy, op. cit., p. 1823.

15. Ibid.

16. Statement of Noel de Nevers, Associate Dean of Engineering, University of Utah, Committee on Interior and Insular Affairs, op. cit., p. 1031.

17. Perry, op. cit., p. 333.

18. Using crude oil prices published in OIL AND GAS JOURNAL, LXX, No. 26 (June 26, 1972), p. 134.

19. Statement of W.W. McClanahan, Jr., National Coal Policy Conference, Inc., Joint Committee on Atomic Energy, op. cit., pp. 1760-1761.

20. U.S. Federal Power Commission, 1971 ANNUAL REPORT (Washington: U.S. Government Printing Office, 1972), p. 45.

21. U.S. Joint Committee on Atomic Energy, NUCLEAR POWER AND RELATED ENERGY PROBLEMS (Washington: U.S. Government Printing Office, 1971), p. 136.

22. Ibid., p. 141.

23. de Nevers, op. cit., p. 1032.

24. Federal Power Commission, 1970 NATIONAL POWER SURVEY, op. cit., p. I-11-18.

25. Shapiro, op. cit., p. 585.

26. Federal Power Commission, 1970 NATIONAL POWER SURVEY, loc. cit.

27. Franklyn C. Rogers, "Underground Nuclear Power Plants: Environmental and Economic Aspects," NUCLEAR NEWS, XIV, No. 5 (May, 1971), p. 39.

28. Statement of A.E. Schubert, Nuclear Technology Division, General Electric Company, Joint Committee on Atomic Energy, HEARINGS, ENVIRONMENTAL EFFECTS OF PRODUCING ELECTRIC POWER, op. cit., p. 1969.

29. Ibid.

30. Federal Power Commission, 1971 ANNUAL REPORT, op. cit., p. 16.

31. Advisory Committee on Underground Transmission, UNDERGROUND POWER TRANSMISSION (Washington: Federal Power Commission, 1966), p. 41.

32. Ibid.

33. Distribution Technical Advisory Committee, THE DISTRIBUTION OF ELECTRIC POWER, reprinted in Federal Power Commission, 1970 NATIONAL POWER SURVEY, op. cit., p. IV-3-73.

34. Ibid., p. IV-3-74.

35. Ibid., p. I-22-8.

Chapter 6
Summary, Conclusions, and Recommendations

1. Statement of James R. Smith, Assistant Secretary for Water and Power Resources, Department of the Interior, U.S. Congress, House, Subcommittee of the Committee on Appropriations, HEARINGS, PUBLIC WORKS FOR WATER AND POWER DEVELOPMENT AND ATOMIC ENERGY COMMISSION APPROPRIATION BILL, 1973, 92nd Cong., 2nd Sess., 1972, pp. 126-128.

2. Statement of Edward E. David, Director, Office of Science and Technology, U.S. Congress, House, Subcommittee on Communications and Power, HEARINGS, POWERPLANT SITING AND ENVIRONMENTAL PROTECTION, 92nd Cong., 1st Sess., 1971, p. 295.

3. Statement of John N. Nassikas, Chairman, Federal Power Commission, Subcommittee of the Committee on Appropriations, op. cit., p. 976.

4. See U.S. Federal Power Commission, Office of Economics, PROMOTIONAL PRACTICES OF PUBLIC UTILITIES: A SURVEY OF RECENT ACTIONS BY STATE REGULATORY COMMISSIONS (Washington: Federal Power Commission, 1970).

Bibliography

Bibliography

Books

Barkley, Paul W., and Seckler, David W. ECONOMIC GROWTH AND ENVIRONMENTAL DECAY. New York: Harcourt Brace Jovanovich, Inc., 1972.

Olson, Charles E. COST CONSIDERATIONS FOR EFFICIENT ELECTRICITY SUPPLY. East Lansing: Michigan State University Press, 1970.

Resources for the Future. U.S. ENERGY POLICIES: AN AGENDA FOR RESEARCH. Baltimore: The Johns Hopkins Press, 1968.

U.S. Atomic Energy Commission. COST-BENEFIT ANALYSIS OF THE U.S. BREEDER REACTOR PROGRAM. Washington: U.S. Government Printing Office, 1972.

_______. FORECAST OF GROWTH OF NUCLEAR POWER, 1972-1985. Washington: U.S. Government Printing Office, 1971.

_______. THE GROWTH OF NUCLEAR POWER, 1972-1985. Washington: U.S. Government Printing Office, 1972.

_______. THE NUCLEAR INDUSTRY-1971. Washington: U.S. Government Printing Office, 1972.

_______. NUCLEAR POWER AND THE ENVIRONMENT. Washington: U.S. Government Printing Office, 1969.

_______. THERMAL EFFECTS AND U.S. NUCLEAR POWER STATIONS. Washington: U.S. Government Printing Office, 1971.

_______. WHY FUSION? Washington: U.S. Government Printing Office, 1970.

U.S. Council on Environmental Quality. THE ECONOMIC IMPACT OF POLLUTION CONTROL. Washington: U.S. Government Printing Office, 1972.

_______. THE PRESIDENT'S 1971 ENVIRONMENTAL PROGRAM. Washington: U.S. Government Printing Office, 1971.

U.S. Department of Health, Education, and Welfare. AIR QUALITY CRITERIA FOR SULFUR OXIDES. Washington: U.S. Government Printing Office, 1969.

_______. CONTROL TECHNIQUES FOR SULFUR OXIDE AIR POLLUTANTS. Washington: U.S. Government Printing Office, 1969.

_______. DANGER IN THE AIR: SULFUR OXIDES AND PARTICULATES. Washington: U.S. Government Printing Office, 1970.

U.S. Environmental Protection Agency. THERMOELECTRIC GENERATORS POWERED BY THERMAL WASTE FROM ELECTRIC POWER PLANTS. Washington: U.S. Government Printing Office, 1970.

U.S. Federal Power Commission. HYDROELECTRIC POWER EVALUATION. Washington: U.S. Government Printing Office, 1968.

_______. HYDROELECTRIC POWER RESOURCES OF THE UNITED STATES. Washington: U.S. Government Printing Office, 1968.

U.S. Federal Power Commission. 1970 NATIONAL POWER SURVEY. Washington: Government Printing Office, 1971.

_______. 1971 ANNUAL REPORT. Washington: U.S. Government Printing Office, 1972.

U.S. Legislative Reference Service. HOW CAN OUR PHYSICAL ENVIRONMENT BEST BE CONTROLLED AND DEVELOPED? Washington: U.S. Government Printing Office, 1970.

U.S. Office of Science and Technology. CONSIDERATIONS AFFECTING STEAM POWER PLANT SITE SELECTION. Washington: U.S. Government Printing Office, 1968.

_______. ELECTRIC POWER AND THE ENVIRONMENT. Washington: U.S. Government Printing Office, 1970.

Articles

Barnea, Joseph. "Geothermal Power," SCIENTIFIC AMERICAN, CCXXVI, No. 1 (January, 1972), pp. 70-77.

Benedict, Manson. "Electric Power from Nuclear Fission," SCIENCE AND PUBLIC AFFAIRS, XXVII, No. 7 (September, 1971), pp. 8-16.

Craig, Roy. "Cloud on the Desert," ENVIRONMENT, XIII, No. 6, (July-August, 1971), pp. 20-35.

Dieckamp, H. "The Fast Breeder Reactor," AMERICAN GEOPHYSICAL UNION TRANSACTIONS, LII, No. 11 (November, 1971), pp. 756-762.

Eliassen, Rolf. "Power Generation and the Environment," SCIENCE AND PUBLIC AFFAIRS, XXVII, No. 7 (September, 1971), pp. 37-42.

Fenner, David, and Klarman, Joseph. "Power from the Earth." ENVIRONMENT, XII, No. 10 (December, 1971), pp. 19-34.

Forbes, Ian A., Ford, Daniel F., Kendall, Henry W., and MacKenzie, James J. "Nuclear Reactor Safety: An Evaluation of New Evidence," NUCLEAR NEWS, XIV, No. 9 (September, 1971), pp. 32-40.

Gambs, Gerard C., and Rauth, Arthur A. "The Energy Crisis, CHEMICAL ENGINEERING, LXXVIII, No. 12 (May 31, 1971), pp. 56-68.

Gregory, Derek P. "A New Concept in Energy Transmission," PUBLIC UTILITIES FORTNIGHTLY, LXXXIX, No. 3 (February 3, 1972), pp. 21-29.

Inglis, David R. "Nuclear Energy and the Malthusian Dilemma," SCIENCE AND PUBLIC AFFAIRS, XXVII, No. 2 (February, 1971), pp. 14-18.

Lear, John. "Radioactive Ashes in the Kansas Salt Cellar," SATURDAY REVIEW, LV, No. 8 (February 19, 1972), pp. 39-42.

Levin, Arthur A., Birch, Thomas J., Hillman, Raines, Gilbert E. "Thermal Discharges: Ecological Effects," ENVIRONMENTAL SCIENCE AND TECHNOLOGY, VI, No. 3 (March, 1972), pp. 224-230.

Lindop, Patricia J., and Rotblat, J. "Radiation Pollution of the Environment,"

SCIENCE AND PUBLIC AFFAIRS, XXVII, No. 7 (September, 1971), pp. 8-16.

McTague, P.J., Davidson, G.J., Bredin, R.M., and Herman, A.A. "The Evolution of Nuclear Plant Costs," NUCLEAR NEWS, XV, No. 2 (February, 1972), pp. 31-35.

Murphy, P.M., Palmer, R.S., and Sherer, D.B. "Incentives for the Development of the Fast Breeder Reactor," PUBLIC UTILITIES FORTNIGHTLY, LXXXIX, No. 7 (March 30, 1972), pp. 25-30.

Post, R.F. "Fusion Power: The Uncertain Certainty," SCIENCE AND PUBLIC AFFAIRS, XXVII, No. 8 (October, 1971), pp. 42-48.

Risser, Hubert E. "Fossil Fuel Supplies and Future Energy Needs," AMERICAN GEOPHYSICAL UNION TRANSACTIONS, LII, No. 11 (November, 1971), pp. 763-767.

Rogers, Franklyn C. "Underground Nuclear Power Plants: Environmental and Economic Aspects," NUCLEAR NEWS, XIV, No. 5 (May, 1971), pp. 36-39.

Snowden, Donald P. "Superconductors for Power Transmission," SCIENTIFIC AMERICAN, CCXXVI, No. 4 (April, 1972), pp. 84-91.

Public Documents

U.S. Congress, Joint Committee on Atomic Energy. HEARINGS ON ENVIRONMENTAL EFFECTS OF PRODUCING ELECTRIC POWER. 91st Cong., 2nd Sess., 1969.

U.S. Congress, House, Subcommittee of the Committee on Appropriations. HEARINGS ON PUBLIC WORKS FOR WATER AND POWER DEVELOPMENT AND ATOMIC ENERGY COMMISSION APPROPRIATIONS BILL, 1973. 92nd Cong., 2nd Sess., 1972.

U.S. Congress, House, Subcommittee on Communications and Power. HEARINGS ON POWERPLANT SITING AND ENVIRONMENTAL PROTECTION. 92nd Cong., 1st Sess., 1971.

U.S. Congress, Senate, Committee on Interior and Insular Affairs. HEARINGS ON PROBLEMS OF ELECTRICAL PRODUCTION IN THE SOUTHWEST. 92nd Cong., 1st Sess., 1971.

U.S. Congress, Senate, Subcommittee on Air and Water Pollution. HEARINGS ON AIR POLLUTION-1970. 91st Cong., 2nd Sess., 1970.

U.S. Congress, Senate, Subcommittee on Air and Water Pollution. HEARINGS ON THERMAL POLLUTION-1968. 90th Cong., 2nd Sess., 1972.

U.S. Congress, Senate, Subcommittee on Intergovernmental Relations. HEARINGS ON INTERGOVERNMENTAL COORDINATION OF POWER DEVELOPMENT AND ENVIRONMENTAL PROTECTION ACT. 91st Cong., 2nd Sess., 1970.

Index

About the Author

David L. Scott is an Assistant Professor of Finance at Florida Southern College, Lakeland, Florida. Born in Rushville, Indiana, he attended Purdue University, Florida State University, and earned the Ph.D. in economics from the University of Arkansas. Professor Scott has been a discussant and author of articles in the field of economics and pollution abatement. In addition, he has contributed articles on financing of education.